BIOFUELS IN THE ENERGY SUPPLY SYSTEM

BIOFUELS IN THE ENERGY SUPPLY SYSTEM

VICTOR I. WELBORNE
EDITOR

Novinka Books
New York

For permission to use material from this book please contact us:
Telephone 631-231-7269; Fax 631-231-8175
Web Site: http://www.novapublishers.com

NOTICE TO THE READER

The Publisher has taken reasonable care in the preparation of this book, but makes no expressed or implied warranty of any kind and assumes no responsibility for any errors or omissions. No liability is assumed for incidental or consequential damages in connection with or arising out of information contained in this book. The Publisher shall not be liable for any special, consequential, or exemplary damages resulting, in whole or in part, from the readers' use of, or reliance upon, this material.

This publication is designed to provide accurate and authoritative information with regard to the subject matter covered herein. It is sold with the clear understanding that the Publisher is not engaged in rendering legal or any other professional services. If legal or any other expert assistance is required, the services of a competent person should be sought. FROM A DECLARATION OF PARTICIPANTS JOINTLY ADOPTED BY A COMMITTEE OF THE AMERICAN BAR ASSOCIATION AND A COMMITTEE OF PUBLISHERS.

LIBRARY OF CONGRESS CATALOGING-IN-PUBLICATION DATA
Available upon request

ISBN 1-59454-756-4

Published by Nova Science Publishers, Inc. ✦ New York

CONTENTS

PREFACE

Biofuels are enjoying a boom throughout the world. Led by ethanol and biodiesel, they are becoming an increasingly popular source of fuel for planes and automobiles. Brazil is the world's leading producer followed by America. Biofuels are far cleaner than petroleum and renewable by plant growth.

Chapter 1 will discuss and compare agriculture-based energy production of ethanol, biodiesel, and wind energy based on three criteria:

- Economic Efficiency compares the price of agriculture-based renewable energy with the price of competing energy sources, primarily fossil fuels.
- Energy Efficiency compares the energy output from agriculture-based renewable energy relative to the fossil energy used to produce it.
- Long-Run Supply Issues consider supply and demand factors that are likely to influence the growth of agriculture-based energy production.

Biofuels are liquid transportation fuels made from plant matter instead of petroleum. Ethanol and biodiesel — the primary biofuels today — can be blended with or directly substitute for gasoline and diesel, respectively. Biofuels reduce air toxics emissions, greenhouse gas buildup, dependence on imported oil, and trade deficits, and support agricultural and rural economies. Chapter 2 -7 report on different aspects of biofuels.

Chapter 8 provides background concerning various aspects of fuel ethanol, and a discussion of the current related policy issues.

In: Biofuels in the Energy Supply System   ISBN: 1-59454-756-4
Editor: Victor I. Welborne, pp. 1-38   © 2006 Nova Science Publishers, Inc.

Chapter 1

AGRICULTURE-BASED RENEWABLE ENERGY PRODUCTION [*]

Randy Schnepf

INTRODUCTION

Agriculture's role as a consumer of energy is well known.[1] However, under the encouragement of expanding government support the U.S. agricultural sector also is developing a capacity to produce energy, primarily as renewable biofuels and wind power. Farm-based energy production has grown rapidly in recent years, but still remains small relative to total national energy needs. In 2003, it provided 0.4% of total U.S. energy consumption (see **Table 1**). Ethanol accounted for about 70% of agriculture-based energy production in 2003; wind energy systems for 29%; and biodiesel energy output for 1%.

In general, fossil-fuel-based energy is less expensive to produce and use than energy from renewable sources.[2] However, since the late 1970s, U.S. policy makers at both the federal and state levels have enacted a variety of incentives, regulations, and programs to encourage the production and use of cleaner, renewable agriculture-based energy. These programs have proven critical to the economic success of rural renewable energy production. The benefits to rural economies and to the environment contrast with the generally higher costs, and have led to numerous proponents as well as

---

[*] Excerpted from CRS Report RL32712 dated January 4, 2005.

critics of the government subsidies that underwrite agriculture-based renewable energy production.

Proponents of government support for agriculture-based renewable energy have cited national energy security, reduction in greenhouse gas emissions, and raising domestic demand for U.S.-produced farm products as viable justification.[3] In addition, proponents argue that rural, agriculture-based energy production can enhance rural incomes and employment opportunities, while encouraging greater value-added for U.S. agricultural commodities.[4]

In contrast, petroleum industry critics of biofuel subsidies argue that technological advances such as seismography, drilling, and extraction continue to expand the fossil-fuel resource base, which remains far cheaper and more accessible than biofuel supplies. Other critics argue that current biofuel production strategies can only be economically competitive with existing fossil fuels in the absence of subsidies if significant improvements in existing technologies are made or new technologies are developed.[5] Until such technological breakthroughs are achieved, critics contend that the subsidies distort energy market incentives and divert research funds from the development of other potential renewable energy sources, such as solar or geothermal, that offer potentially cleaner, more bountiful alternatives.

Still others question the rationale behind policies that promote biofuels for energy security. These critics question whether the United States could ever produce sufficient feedstocks of either starches, sugars, or vegetable oils to permit biofuel production to meaningfully offset petroleum imports.[6] Finally, there are those who argue that the focus on development of alternative energy sources undermines efforts to conserve and reduce the nation's energy dependence.

This article will discuss and compare agriculture-based energy production of ethanol, biodiesel, and wind energy based on three criteria:

- **Economic Efficiency** compares the price of agriculture-based renewable energy with the price of competing energy sources, primarily fossil fuels.
- **Energy Efficiency** compares the energy output from agriculture-based renewable energy relative to the fossil energy used to produce it.
- **Long-Run Supply Issues** consider supply and demand factors that are likely to influence the growth of agriculture-based energy production.

Several additional criteria may be used for comparing different fuels, including performance, emissions, safety, and infrastructure needs. For more information on these additional criteria and others, see the Department of Energy (DOE), Energy Efficiency and Renewable Energy (EERE), Alternative Fuels Data Center, at [http://www.eere.energy.gov/afdc/altfuel/fuel_properties.html].

AGRICULTURE'S SHARE OF ENERGY PRODUCTION

In 2003, the major agriculture-produced energy source — ethanol — accounted for about 0.8% of U.S. petroleum consumption and about 0.3% of total U.S. energy consumption (see **Table 1**). In addition to ethanol production, several other renewable energy sources — biodiesel, wind, anaerobic digesters, and non-traditional biomass — also appear to offer particular advantages to the agricultural sector. Presently, the volume of agriculture-based energy produced from these emerging renewable sources is small relative to ethanol production. However, an expanding list of federal and state incentives, regulations, and programs that were enacted over the past decade have helped to encourage more diversity in renewable energy production and use.[7]

Agriculture-Based Biofuels

Biofuels are liquid fuels produced from biomass. Types of biofuels include ethanol, biodiesel, methanol, and reformulated gasoline components.[8] The Biomass Research and Development Act of 2000 (P.L. 106-224; Title III) defines biomass as "any organic matter that is available on a renewable or recurring basis, including agricultural crops and trees, wood and wood wastes and residues, plants (including aquatic plants), grasses, residues, fibers, and animal wastes, municipal wastes, and other waste materials."

Table 1. U.S. Energy Production and Consumption, 2003

	Production		Consumption	
	Quadrillion	% of	Quadrillion	% of
Energy source	Btu	total	Btu	total
Total	70.5	100.0%	98.2	100.0%
Fossil Fuels	56.4	80.1%	84.4	85.9%
Nuclear	8.0	11.3%	8.0	8.1%
Renewables	6.2	8.7%	6.2	6.3%
Fossil Fuel Categories				
Petroleum and products	12.1	17.2%	39.0	39.8%
Coal	22.3	31.7%	22.7	23.1%
Natural Gas	22.0	31.2%	22.6	23.0%
Renewable Categories				
Hydroelectric power	2.8	3.9%	2.8	2.8%
Wood, waste, oth. alcohol	2.6	3.7%	2.6	2.7%
Geothermal	0.3	0.4%	0.3	0.3%
Solar	0.1	0.1%	0.1	0.1%
Ethanol	0.3	0.4%	0.3	0.3%
Biodiesel	0.0	0.0%	0.0	0.0%
Wind	0.1	0.2%	0.1	0.1%

Source: Ethanol data: American Coalition on Ethanol, [http://www.ethanol.org]; biodiesel data: National Biodiesel Board, [http://www.biodiesel.org]; all other data: DOE, Energy Information Agency (EIA), Table 1.2, "Energy Production by Source, 1949-2003," and Table 1.3, "Total U.S. Energy Consumption by Source."

Biofuels are primarily used as transportation fuels for cars, trucks, buses, airplanes, and trains. As a result, their principal competitors are gasoline and diesel fuel. Unlike fossil fuels, which have a fixed resource base that declines with use, biofuels are produced from renewable feedstocks. Furthermore, under most circumstances biofuels are more environmentally friendly (in terms of emissions of toxins, volatile organic compounds, and greenhouse gases) than petroleum products. Supporters of biofuels emphasize that biofuel plants generate value-added economic activity that increases demand for local feedstocks, which raises commodity prices, farm incomes, and rural employment.

Ethanol[9]

Ethanol, or ethyl alcohol, is an alcohol made by fermenting and distilling simple sugars. As a result, ethanol can be produced from any biological feedstock that contains appreciable amounts of sugar or materials that can be converted into sugar such as starch or cellulose. Sugar beets and sugar cane are examples of feedstocks that contain sugar. Corn contains starch that can relatively easily be converted into sugar. In the United States corn is the principal ingredient used in the production of ethanol; in Brazil (the world's largest ethanol producer), sugar cane is the primary feedstock. A significant percentage of trees and grasses are made up of cellulose which can also be converted to sugar, although with more difficulty than required to convert starch. In recent years, researchers have begun experimenting with the possibility of growing hybrid grass and tree crops explicitly for ethanol production. In addition, sorghum and potatoes, as well as crop residue and animal waste, are potential feedstocks.

Ethanol production has shown rapid growth in the United States in recent years (see **Figure 1**). Several events have contributed to greater ethanol production: the energy crises of the early and late 1970s, a partial exemption from the motor fuels excise tax (legislated as part of the Energy Tax Act of 1978), ethanol's emergence as a gasoline oxygenate, and provisions of the Clean Air Act Amendments of 1990 that favored ethanol blending with gasoline.[10]

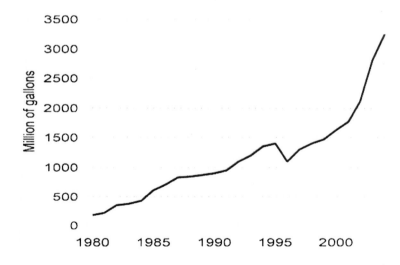

Figure 1. U.S. Ethanol Production, Annual 1980-2004

In November 2004, existing U.S. ethanol plant capacity was a reported 3,419 million gallons per year, with an additional capacity of 706 million gallons per year under construction. U.S. ethanol production presently is underway or planned in 25 states centered around the central and western Corn Belt, where corn supplies are most plentiful (see **Table 2**).[11] Corn accounts for over 90% of the feedstocks used in ethanol production in the United States.

Table 2. Ethanol Production Capacity by State, November 2004

State	Operating	Under Construction	Total
		Million gallons per year	
Iowa	872	350	1,222
Illinois	695	33	728
Minnesota	398	149	546
Nebraska	492	0	492
South Dakota	384	20	404
Kansas	134	20	154
Wisconsin	87	40	127
Missouri	65	40	105
Indiana	95	0	95
Others	197	54	251
U.S. Total	**3,419**	**706**	**4,124**

Source: Renewable Fuels Association, *U.S. Fuel Ethanol Production Capacity*, at [http://www.ethanolrfa.org/eth_prod_fac.html], November 2004.

Corn-Based Ethanol. USDA projects that 1.425 billion bushels of corn (or 12.2% of total U.S. corn production) will be used from the 2004 corn crop to produce up to 3.8 billion gallons of ethanol during 2004/05 (September-August).[12] In gasoline-equivalent gallons (GEG), this represents more than 2.5 billion gallons.[13] Despite its rapid growth, ethanol production represents a minor part of U.S. gasoline consumption, with a projected 1.6% share in 2004 (2.1 billion GEG out of 136.4 billion gallons of total gasoline use).[14]

Economic Efficiency. Ethanol's primary fuel competitor is gasoline. Wholesale ethanol prices, before incentives from the federal and state governments, are generally significantly higher than those of their fossil fuel counterparts. For example, during the week of June 14, 2004, the average retail price of E85 (a blend of 85% ethanol with 15% gasoline) ranged between $2.28 and $2.70 per GEG, compared with a range of $1.92 to $2.24

for regular grade gasoline (see **Table 3**).[15]The approximate price difference of 36¢ to 46¢ implies that pure ethanol costs as much as 42¢ to 54¢ per GEG more than gasoline. The federal production tax credit of 51¢ per gallon of pure ethanol (see below) offsets most of the price difference, thereby helping ethanol to compete in the marketplace.

Apart from government incentives, the economics underlying corn-based ethanol's market competitiveness hinge on the following factors:

- the price of feedstocks, primarily corn;
- the price of the processing fuel, primarily natural gas or electricity, used at the ethanol plant;
- the cost of transporting feedstocks to the ethanol plant and transporting the finished ethanol to the user; and
- the price of feedstock co-products (for dry-milled corn: distillers dried grains; for wet-milled corn: corn gluten feed, corn gluten meal, and corn oil).

Higher prices for corn, processing fuel, and transportation hurt ethanol's market competitiveness, while higher prices for corn by-products and gasoline improve ethanol's competitiveness in the marketplace. Feedstock costs are the largest single cost factor in the production of ethanol. As a result, the relative relationship of corn to gasoline prices provides a strong indicator of the ethanol industry's well-being. A comparison of corn versus gasoline prices (see **Figure 2**) suggests that the general trend since the late 1990s has clearly been in ethanol's favor as national average monthly gasoline prices have surged towards the $2.00 per gallon level while corn prices have returned to the $2.00 per bushel level of the early 2000s.

Government Support. Federal subsidies help ethanol to overcome its higher cost relative to gasoline. The primary federal incentives include:[16]

- a production tax credit of 51¢ per gallon of pure (100%) ethanol — the tax incentive was extended through 2010 and converted to a tax credit from a partial tax exemption of the federal excise tax under the American Jobs Creation Act of 2004 (P.L. 108-357);
- a small producer income tax credit (26 USC 40) of 10¢ per gallon for the first 15 million gallons of production for ethanol producers whose total output does not exceed 30 million gallons of ethanol per year; and

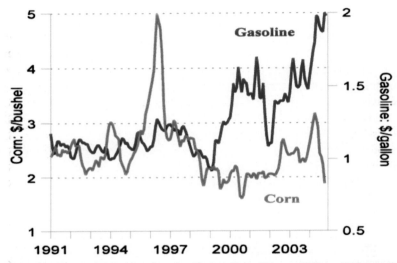

Source: Prices are monthly averages: Corn, No.2, yellow, Chicago; USDA, AMS; gasoline
prices are national average retail, DOE, EIA.

Figure 2. Corn versus Gasoline Prices, 1991-2004

- incentive payments (contingent on annual appropriations) on year-to-year production increases of renewable energy under USDA's Bioenergy Program (7 U.S.C. 8108).

Indirectly, other federal programs support ethanol production by requiring federal agencies to give preference to biobased products in purchasing fuels and other supplies and by providing incentives for research on renewable fuels. Also, several states have their own incentives, regulations, and programs in support of renewable fuel research, production, and consumption that supplement or exceed federal incentives.

Energy Efficiency. The net energy balance (NEB) of a fuel can be expressed as a ratio of the energy produced from a production process relative to the energy used in the production process. An output/input ratio of 1.0 implies that energy output equals energy input. The critical factors underlying ethanol's energy efficiency or NEB include:

- corn yields;
- the energy efficiency of corn production, including the energy embodied in inputs such as fertilizers, pesticides, seed corn, and cultivation practices;

Table 3. Energy and Price Comparisons: Various Fuels

Fuel type	Unit	Btu's per unit[a]	Average Price: $ per unit	GEG[b]	Average Price: $ per GEG[c]
Gasoline:					
conventional	gallon	125,071	$1.92 - $2.24	1.00	$1.92 - $2.24
Ethanol (E85)	gallon	83,361	$1.82 - $2.16	0.80	$2.28 - $2.70
Diesel fuel	gallon	138,690	$1.64 - $2.00	1.11	$1.48 - $1.80
Biodiesel (B20)	gallon	138,690	$1.73 - $2.11	1.11	$1.56 - $1.90
Propane	gallon	91,333	$1.40 - $2.25	0.74	$1.88 - $3.02
Natural Gas[d]	1,000 ft.^{3}	1,030	$6.08 - $8.08	na	$1.02 - $1.54
Biogas	1,000 ft.^{3}	10 x (% methane)[e]	na	na	na
Electricity[f]	kilowatt-hour	3,413	5¢ - 16¢	na	na

Source: Conversion rates for petroleum-based fuels and electricity are from DOE, *Monthly Energy Review*, August 2004. na = not applicable. [a] A Btu (British thermal unit) is a measure of the heat content of a fuel and indicates the amount of energy contained in the fuel. Because energy sources vary by form (gas, liquid, or solid) and energy content, the use of Btu's allows the adding of various types of energy using a common benchmark. [b] GEG = gasoline equivalent gallon. The GEG allows for comparison across different forms — gas, liquid, kilowatt, etc. It is derived from the Btu content by first converting each fuel's units to gallons, then dividing each fuel's Btu unit rate by gasoline's Btu unit rate of 125,000, and finally multiplying each fuel's volume by the resulting ratio. [c] DOE, EIA, *The Alternative Fuel Price Report*, June 29, 2004. Prices are for mid-June 2004. The retail price per gallon has been converted to price per GEG units. [d] Natural Gas prices, $ per 1,000 cu. ft., are industrial prices for the month of July 4, 2004, from DOE, EIA, available at [http://tonto.eia.doe.gov/dnav/ng/ng_pri_sum_dcu_nus_m.htm]. [e] When burned, biogas yields about 10 Btu per percentage of methane composition. For example, 65% methane yields 650 Btu per 1,000 cubic foot. [f] Prices are for commercial electricity rates per kilowatt-hour.

- the energy efficiency of the corn-to-ethanol production process — about 55% of the corn used for ethanol is processed by "dry" milling (a grinding process); 45% is processed by "wet" milling plants (a chemical extraction process); and
- the energy value of corn by-products.[17]

Over the past decade technical improvements in the production of agricultural inputs (particularly nitrogen fertilizer) and ethanol, coupled with higher corn yields per acre and stable or lower input needs, appear to have raised ethanol's NEB. In 2004, USDA economists reported that, assuming "best production practices and state of the art processing technology," the NEB of corn-ethanol (based on 2001 data) was a positive 1.67 — that is, 67% more energy was returned from a gallon of ethanol than was used in its production.[18] This compares with an NEB of 0.81 for gasoline — that is, 19% less energy is returned from a gallon of gasoline than is used in its life cycle from source to user.[19] Other researchers have found much lower NEB values under less optimistic assumptions.[20]

Long-Run Supply Issues. Despite improving energy efficiency, the ability for domestic ethanol production to measureably substitute for petroleum imports is questionable, particularly when ethanol production depends almost entirely on corn as the primary feedstock. The U.S. petroleum import share is estimated at 54% of consumption in 2004 and is expected to grow to a 70% share by 2025.[21] Presently, ethanol production accounts for about 1.6% of U.S. gasoline consumption while using about 12% of the U.S. corn production. If the entire 2004 U.S. corn production were dedicated to ethanol production, the resultant 20.3 billion GEG would represent about 15% of projected national gasoline use of 136.4 billion gallons.[22] In 2004, slightly more than 73 million acres of corn are expected to be harvested. At least 140 million acres would be needed to produce enough corn and subsequent ethanol to substitute for 50% of gasoline imports (or 27% of U.S. consumption).[23] Since 1970, corn harvested acres have never reached 76 million acres. Thus, barring a drastic realignment of U.S. field crop production patterns, corn-based ethanol's potential as a petroleum import substitute appears to be limited by a crop area constraint.

Domestic and international demand places additional limitations on corn use for ethanol production in the United States. Corn traditionally represents about 57% of feed concentrates and processed feedstuffs fed to animals in the United States.[24]Also, the United States is the world's leading corn exporter, with nearly a 66% share of world trade during the past decade. In 2003/04, the United States exported nearly 19% of its corn production.

Plus, there is an inherent tradeoff in using a widely consumed agricultural product for a non-agricultural use. As corn-based ethanol production increases, so does total corn demand and corn prices. Higher corn prices, in turn, mean higher feed costs for cattle, hog, and poultry producers. The corn co-products from ethanol processing would likely substitute for some of the lost feed value of corn used in ethanol processing.[25] However, about 66% of the original weight of corn is consumed in producing ethanol and is no longer available for feed.[26] Higher corn prices would also likely result in lost export sales. International feed markets are very price sensitive as several different grains and feedstuffs are relatively close substitutes. A sharp rise in U.S. corn prices would likely result in a more than proportionate decline in corn exports.

Furthermore, as ethanol production increases, the energy (derived primarily from natural gas) needed to process the corn into ethanol would increase. For example, the energy needed to process the entire 2004 corn crop into ethanol would be approximately 1.6 trillion cubic feet of natural gas (or 1.4 trillion cu. ft. more than is currently used).[27] Total U.S. natural gas consumption was about 22.6 trillion cu.ft. in 2003. The United States has been a net importer of natural gas since the early 1980s. Because natural gas is used extensively in electricity production in the United States, significant increases in its use as a processing fuel in the production of ethanol would likely result in substantial increases in both prices and imports of natural gas.

These supply issues suggest that corn's long-run potential as an ethanol feedstock is somewhat limited. According to the DOE, the cost of producing and transporting ethanol will continue to limit its use as a renewable fuel; ethanol relies heavily on federal and state support to remain economically viable; and the supply of ethanol is extremely sensitive to corn prices, as seen in 1996 when record farm prices received for corn led to a sharp reduction in U.S. ethanol production. Finally, DOE suggests that the ability to produce ethanol from low-cost biomass will ultimately be the key to making it competitive as a gasoline additive.[28]

In contrast to expanded biofuel production, research suggests that far greater fuel economies could be obtained by a small adjustment in existing vehicle mileage requirements. For example, an increase in fuel economy of one mile per gallon across all passenger vehicles in the United States could cut petroleum consumption by more than all alternative fuels and replacement fuels combined.[29]

Ethanol from Cellulosic Biomass Crops.[30] Besides corn, several other agricultural products are viable feedstocks and appear to offer long-term supply potential — particularly cellulose-based feedstocks. An emerging cellulosic feedstock with apparently large potential as an ethanol feedstock is switchgrass, a native grass that thrives on marginal lands as well as on prime cropland, and needs little water and no fertilizer. The opening of Conservation Reserve Program (CRP) land to switchgrass production under Section 2101 of the 2002 farm bill (P.L. 107-171) has helped to spur interest in its use as a cellulosic feedstock for ethanol production. Other potential cellulose-to-ethanol feedstocks include fast-growing woody crops such as hybrid poplar and willow trees, as well as waste biomass materials —

logging residues, wood processing mill residues, urban wood wastes, and selected agricultural residues such as sugar cane bagasse and rice straw.

The main impediment to the development of a cellulose-based ethanol industry is the state of cellulosic conversion technology (i.e., the process of converting cellulose-based feedstocks into fermentable sugars). Currently, cellulosic conversion technology is rudimentary and expensive. As a result, no commercial cellulose-to-ethanol facilities are in operation in the United States, although plans to build several facilities are underway. On April 21, 2004, Iogen — a Canadian firm — became the first firm to successfully engage in the commercial production of cellulosic ethanol (from wheat straw) at a large-scale demonstration plant in Ottawa.[31] In addition, pilot facilities are operational in both the United States and Canada.

Economic Efficiency. The conversion of cellulosic feedstocks to ethanol parallels the corn conversion process, except that the cellulose must first be converted to fermentable sugars. As a result, the key factors underlying cellulosic-based ethanol's price competitiveness are essentially the same as for corn-based ethanol, with the addition of the cost of cellulosic conversion.

Cellulosic feedstocks are significantly less expensive than corn; however, at present they are more costly to convert to ethanol because of the extensive processing required. Currently, cellulosic conversion is done using either dilute or concentrated acid hydrolysis — both processes are prohibitively expensive. However, the DOE suggests that enzymatic hydrolysis, which processes cellulose into sugar using cellulase enzymes, offers both processing advantages as well as the greatest potential for cost reductions. Current cost estimates of cellulase enzymes range from 30¢ to 50¢ per gallon of ethanol.[32] The DOE is also studying thermal hydrolysis as a potentially more cost-effective method for processing cellulose into sugar.

Based on the state of existing technologies and their potential for improvement, the DOE estimates that improvements to enzymatic hydrolysis could eventually bring the cost to less than 5¢ per gallon, but this may still be a decade or more away. Were this to happen, then the significantly lower cost of cellulosic feedstocks would make cellulosic-based ethanol dramatically less expensive than corn-based ethanol and gasoline at current prices.

Iogen's breakthrough involved the successful use of recombinant DNA-produced enzymes to break apart cellulose to produce sugar for fermentation into ethanol. Both the DOE and USDA are funding research to improve cellulosic conversion as well as to breed higher yielding cellulosic crops. In

1978, the DOE established the Bioenergy Feedstock Development Program (BFDP) at the Oak Ridge National Laboratory. The BFDP is engaged in the development of new crops and cropping systems that can be used as dedicated bioenergy feedstocks. Some of the crops showing good cellulosic production per acre with strong potential for further gains include fast-growing trees (e.g., hybrid poplars and willows), shrubs, and grasses (e.g., switchgrass).

Government Support. Although no commercial cellulosic ethanol production has occurred yet in the United States, two provisions of the 2002 farm bill (P.L. 107-171) have encouraged research in this area. The first provision (Section 2101) allows for the use of Conservation Reserve Program lands for wind energy generation and biomass harvesting for energy production and has helped to spur interest in hardy biofuel feedstocks that are able to thrive on marginal lands. Another provision (Section 9008) provides competitive funding for research and development projects on biofuels and bio-based chemicals in an attempt to motivate further production and use of non-traditional biomass feedstocks.[33]

Energy Efficiency. The use of cellulosic biomass in the production of ethanol yields a higher net energy balance compared to corn — a 34% net gain for corn vs. a 100% gain for cellulosic biomass — based on a 1999 comparative study.[34]While corn's net energy balance (under optimistic assumptions concerning corn production and ethanol processing technology) has been estimated at 67% by USDA in 2004, it is likely that cellulosic biomass's net energy balance would also have experienced parallel gains for the same reasons — improved crop yields and production practices, and improved processing technology.

Long-Run Supply Issues. Cellulosic feedstocks have an advantage over corn in that they grow well on marginal lands, whereas corn requires fertile cropland (as well as timely water and the addition of soil amendments). This greatly expands the potential area for growing cellulosic feedstocks relative to corn. For example, in 2001 nearly 76 million acres were planted to corn, out of 244 million acres planted to the eight major field crops (corn, soybeans, wheat, cotton, barley, sorghum, oats, and rice). In contrast, that same year the United States had 433 million acres of total cropland (including forage crops and temporarily idled cropland) and 578 million acres of permanent pastureland, most of which is potentially viable for switchgrass production.[35]

A 2003 BFDP study suggests that if 42 million acres of cropped, idle, pasture, and CRP acres were converted to switchgrass production, 188 million dry tons of switchgrass could be produced annually (at an implied yield of 4.5 metric tons per acre), resulting in the production of 16.7 billion gallons of ethanol or 10.9 billion GEG.[36] This would represent about 8% of U.S. gasoline use in 2003. Existing research plots have produced switchgrass yields of 15 dry tons per acre per year, suggesting tremendous long-run production potential. However, before any supply potential can be realized, research must first overcome the cellulosic conversion cost issue through technological developments.

Methane from an Anaerobic Digester[37]

An anaerobic digester is a device that promotes the decomposition of manure or "digestion" of the organics in manure by anaerobic bacteria (in the absence of oxygen) to simple organics while producing biogas as a waste product. The principal components of biogas from this process are methane (60% to 70%), carbon dioxide (30% to 40%), and trace amounts of other gases. Methane is the major component of the natural gas used in many homes for cooking and heating, and is a significant fuel in electricity production. Biogas can also be used as a fuel in a hot water heater if hydrogen sulfide is first removed from the biogas supply. As a result, the generation and use of biogas can significantly reduce the cost of electricity and other farm fuels such as natural gas, propane, and fuel oil.

By late 2002, there were an estimated 40 digester systems in operation at commercial U.S. livestock farms, with an additional 30 expected to be in operation by 2003.[38] Anaerobic digestion system proposals have frequently received funding under the Renewable Energy Program (REP) of the 2002 farm bill (P.L. 107-171, Title IX, Section 9008). In 2004, 37 anaerobic digester proposals from 26 different states were awarded funding under the REP.[39] Also, the AgStar program — a voluntary cooperative effort by USDA, EPA, and DOE — encourages methane recovery at confined livestock operations that manage manure as liquid slurries.[40]

Economic Efficiency. The primary benefits of anaerobic digestion are animal waste management, odor control, nutrient recycling, greenhouse gas reduction, and water quality protection. Except in very large systems, biogas production is a highly useful but secondary benefit. As a result, anaerobic digestion systems do not effectively compete with other renewable energy production systems on the basis of energy production alone. Instead, they compete with and are cost-competitive when compared to conventional waste management practices according to EPA.[41] Depending on the infrastructure design — generally some combination of storage pond, covered or aerated treatment lagoon, heated digester, and open storage tank — anaerobic digestion systems can range in investment cost from $200 to $500 per Animal Unit (i.e., per 1,000 pounds of live weight). In addition to the initial infrastructure investment, recurring costs include manure and effluent handling, and general maintenance. According to EPA, these systems can have financially attractive payback periods of three to seven years when energy gas uses are employed. On average, manure from a lactating 1,400-pound dairy cow can generate enough biogas to produce 550 Kilowatts per year.[42] A 200-head dairy herd could generate 500 to 600 Kilowatts per day. At 6¢ per kWh, this would represent potential energy cost savings of $6,600 per year.

The principal by-product of anaerobic digestion is the effluent (i.e., the digested manure). Because anaerobic digestion substantially reduces ammonia losses, the effluent is more nitrogen-rich than untreated manure, making it more valuable for subsequent field application. Also, digested manure is high in fiber, making it valuable as a high-quality potting soil ingredient or mulch. Other cost savings include lower total lagoon volume requirements for animal waste management systems (which reduces excavation costs and the land area requirement), and lower cover costs because of smaller lagoon surface areas.

Energy Efficiency. Because biogas is essentially a by-product of an animal waste management activity, and because the biogas produced by the system can be used to operate the system, the energy output from an anaerobic digestion system can be viewed as achieving even or positive energy balance. The principal energy input would be the fuel used to operate the manure handling equipment.

Long-Run Supply Issues. Anaerobic digesters are most feasible alongside large confined animal feeding operations (CAFOs). According to USDA, biogas production for generating cost effective electricity requires manure from more than 150 large animals. As animal feeding operations steadily increase in size, the opportunity for anaerobic digestion systems will likewise increase.

Biodiesel

Biodiesel is an alternative diesel fuel that is produced from any animal fat or vegetable oil (such as soybean oil or recycled cooking oil). About 90% of U.S. biodiesel is made from soybean oil. As a result, U.S. soybean

producers and the American Soybean Association (ASA) are strong
advocates for greater government support for biodiesel production.

According to the National Biodiesel Board (NBB), biodiesel is nontoxic,
biodegradable, and essentially free of sulfur and aromatics. In addition, it
works in any diesel engine with few or no modifications and offers similar
fuel economy, horsepower, and torque, but with superior lubricity and
important emission improvements over petroleum diesel.

U.S. biodiesel production has shown strong growth in recent years,
increasing from under 1 million gallons in 1999 to over 30 million gallons in
2004 (see **Figure 3**). However, U.S. biodiesel production remains small
relative to national diesel consumption levels. In 2003, biodiesel production
of 25 million gallons represented 0.06% of the 40 billion gallons of diesel
fuel used nationally for vehicle transportation.[43] In addition to vehicle use,
17.8 billion gallons of diesel fuel were used for heating and power
generation by residential, commercial, and industry, and by railroad and
vessel traffic in 2002, bringing total U.S. diesel fuel use to nearly 58 billion
gallons (see **Table 4**).

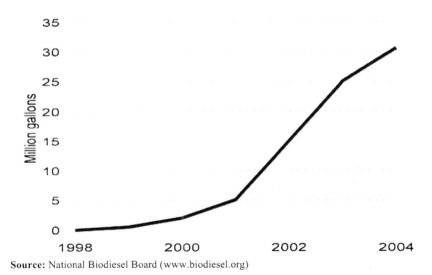

Source: National Biodiesel Board (www.biodiesel.org)

Figure 3. U.S. Biodiesel Production, 1998-2004

As of October 2004, there were more than 20 companies producing and
marketing biodiesel commercially in the United States, and another 20 new
firms that have reported their plans to construct dedicated biodiesel plants in
the near future.[44]The NBB reported that mid-2004 U.S. biodiesel

production capacity was an estimated 150 million gallons per year, but would likely double in size over the next 12-18 months based on the number of ongoing biodiesel projects.[45] But many of these plants also can produce other products, for example, cosmetics, so total capacity (and capacity for expansion) is far greater than actual biodiesel production.

Economic Efficiency. Biodiesel is generally more expensive than its fossil fuel counterpart. For example, during the week of June 14, 2004, the average retail price of B20 (a blend of 20% biodiesel with 80% conventional diesel) ranged from $1.73 to $2.11 per gallon compared with a range of $1.64 to $2.00 for conventional diesel fuel (see **Table 3**).[46] The approximate price difference of 10¢ implies that pure biodiesel costs as much as 50¢ more per gallon to produce.

Table 4. U.S. Diesel Fuel Use, 2002

U.S. Diesel Use in 2002	Total Million gallons[a]	Hypothetical Scenario: 1% of Total Use[b]	
		Million gallons	Soybean Oil Equivalents: Million pounds[a]
Total Vehicle Use	40,043	400	3,083
On-Road	34,309	343	2,642
Off-Road	2,224	22	171
Military	331	3	25
Farm	3,179	32	245
Total Non-vehicle Use	17,842	178	1,374
All Uses	57,885	579	4,457

Source:U.S. diesel use is from DOE, EIA, *U.S. Annual Distillate Sales by End Use.* [a]Pounds are converted from gallons of oil using a 7.7 pounds-to-gallon conversion rate. [b]Hypothetical scenario included for comparison purposes only.

The prices of biodiesel feedstocks, as well as petroleum-based diesel fuel, vary over time based on domestic and international supply and demand conditions (see **Figure 4**). As diesel fuel prices rise relative to biodiesel, and/or as biodiesel production costs fall through lower commodity prices or technological improvements in the production process, biodiesel becomes more economical. In addition, federal and state assistance helps to make biodiesel more competitive with diesel fuel.

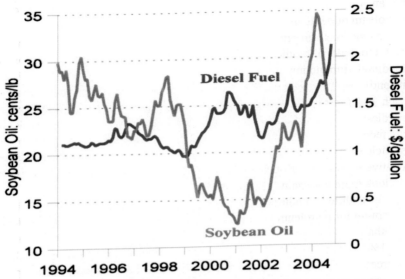

Source: diesel fuel: DOE, EIA; soybean oil; USDA, FAS "Oilseed Circular."

Figure 4. Soybean Oil vs Diesel Fuel, 1994 to 2004

Government Support. The primary federal incentive for biodiesel production is a production excise tax credit signed into law on October 22, 2004, as part of the American Jobs Creation Act of 2004 (P.L. 108-357). Under the biodiesel production tax credit, the subsidy amounts to $1.00 for every gallon of agri-biodiesel (i.e., virgin vegetable oil and animal fat) that is used in blending with petroleum diesel. A 50¢ credit is available for every gallon of non-agri-biodiesel (i.e., recycled oils such as yellow grease). At current prices, the federal tax credit would make biodiesel very competitive with petroleum-based diesel fuel, as the 20¢ tax credit on a gallon of B20 would more than offset the 10¢ price difference with conventional diesel. However, unlike the ethanol tax credit, which was extended through 2010, the biodiesel tax credit expires at the end of calendar year 2006.

In addition to the production tax credit, USDA's Bioenergy Program (7 U.S.C. 8108) provides incentive payments (contingent on annual appropriations) on year-to-year production increases of renewable energy.

Energy Efficiency. Biodiesel appears to have a significantly better net energy balance than ethanol, according to a joint USDA-DOE 1998 study that found biodiesel to have an NEB of 3.2 — that is, 220% more energy was returned from a gallon of pure biodiesel than was used in its production.[47] In contrast, the study authors point out that petroleum diesel has an NEB of 0.83 — that is, 17% less energy was returned from a gallon of petroleum diesel than was used in its life cycle from source to user.

Long-Run Supply Issues. Both the ASA and the NBB are optimistic that the federal biodiesel tax incentive will provide the same boost to biodiesel production that ethanol has obtained from its federal tax incentive.[48] However, many commodity market analysts are skeptical of such claims. They contend that the biodiesel industry still faces several hurdles: the retail distribution network for biodiesel has yet to be established; the federal tax credit, which expires on December 31, 2006, does not provide sufficient time for the industry to develop; and potential oil feedstocks are relatively less abundant than ethanol feedstocks, making the long-run outlook more uncertain.

In addition, biodiesel production confronts the same limited ability to substitute for petroleum imports and the same type of consumption tradeoffs as ethanol production. If, under a hypothetical scenario (as shown in **Table 4**), 1% of current vehicle diesel fuel use were to originate from biodiesel sources, this would require about 400 million gallons of biodiesel (compared to current production of about 30 million gallons) or approximately 3 billion pounds of soybean oil. During 2003, a total of 31.7 billion pounds of vegetable oils and animal fats were produced in the United States; however, most of this production was committed to other food and industrial uses. Uncommitted biodiesel feedstocks (as measured by the available stock levels on September 30, 2003) were 2.1 billion pounds (see **Table 5**).

If biodiesel were to be 1% of diesel use, an additional 900 million pounds of soybean oil would be needed after exhausting all available feedstocks. This is nearly equivalent to the 937 million pounds of soybean oils exported by the United States in 2003/04. If soybean oil exports were to remain unchanged, the deficit feedstocks could be obtained either by reducing U.S. whole soybean exports by about 80 million bushels or by expanding soybean production by approximately 1.6 million acres (assuming a yield of about 50 bushels per acre). A further possibility is that U.S. producers could shift towards the production of higher-oil content oilseeds such as canola or sunflower.

The bottom line is that a small increase in demand of fats and oils for biodiesel production could quickly exhaust available feedstock supplies and push vegetable oil prices significantly higher due to the low elasticity of demand for vegetable oils in food consumption.[49] At the same time, it would begin to disturb feed markets.

Table 5. U.S. Potential Biodiesel Feedstocks, 2002/03

	Wholesale Price[a]	Oil Production, 2002/03		Ending Stocks: Sept. 30, 2003	
Oil type	$/lb	Million pounds	Million gallons[a]	Million pounds	Million gallons[a]
Crops		23,050	2,994	1,834	238
Soybean	20.6	18,435	2,394	1,486	193
Corn	22.3	2,453	319	114	15
Cottonseed	25.7	725	94	40	5
Sunflowerseed	26.4	320	42	25	3
Canola	23.6[c]	541	70	55	7
Peanut	44.5	286	37	50	6
Flaxseed/linseed	na	201	26	45	6
Safflower	na	89	12	19	2
Rapeseed	23.6[c]	0	0	0	0
Animal fat & other		8,698	1,130	299	39
Lard	18.1	262	34	9	1
Edible tallow	16.9	1,974	256	26	3
Inedible tallow	na	3,690	479	221	29
Yellow grease	11.6	2,772	360	43	6
Total supply		31,748	4,123	2,134	277

Source: USDA, ERS, *Oil Crops Yearbook*, OCS-2003, October 2003. Rapeseed was calculated by multiplying oil production by a 40% conversion rate. The inedible tallow and yellow grease supplies come from Dept of Commerce, Bureau of Census, *Fats and Oils, Production, Consumption and Stocks, Annual Summary 2002.* na = not available. [a]Average of monthly price quotes for 2000/01 to 2003/04 marketing years (Oct. to Sept.). USDA, ERS, *Oil Crops Outlook*, various issues. Yellow grease price is a 1993-95 average from USDA, ERS, AER 770, Sept. 1998, p. 9. [b]Pounds are converted to gallons of oil using a 7.7 pounds-to-gallon conversion rate. [c]The price average is for rapeseed oil, f.o.b., Rotterdam; USDA, FAS, *Oilseeds: World Market and Trade*, various issues.

As with ethanol production, increased soybean oil production (dedicated to biodiesel production) would generate substantial increases in animal feeds in the form of high-protein meals. When a bushel of soybeans is processed (or crushed), nearly 80% of the resultant output is in the form of soybean meal, while only about 18%-19% is output as soybean oil. Thus, for every 1 pound of soybean oil produced by crushing whole soybeans, over 4 pounds of soybean meal are also produced.

Crushing an additional 80 million bushels of soybeans for soybean oil would produce over 1.9 million short tons (s.t.) of soybean meal. In 2003/04, the United States produced 36.6 million s.t. of soybean meal. An additional 1.9 million s.t. of soybean meal entering U.S. feed markets would compete directly with the feed byproducts of ethanol production (distillers dried grains, corn gluten feed, and corn gluten meal) with economic ramifications that have not yet been fully explored.[50]Also similar to ethanol production, natural gas demand would likely rise with the increase in biodiesel processing.[51]

WIND ENERGY SYSTEMS

In 2003, electricity from wind energy systems accounted for about 0.1% of U.S. total energy consumption (see **Table 1**). However, wind-generated electricity compared more favorably as a share of electricity used by the U.S. agriculture sector (28%), or of total direct energy used by U.S. agriculture (9%) that same year.[52] Total installed wind energy production capacity has expanded rapidly in the United States in the past six years, rising from 1,848 megawatts (MW)[53] in 1998 to nearly 6,400 MW by 2003 (see **Figure 5**).[54] About 90% of capacity is in 10 predominantly midwestern and western states (see **Table 6**).

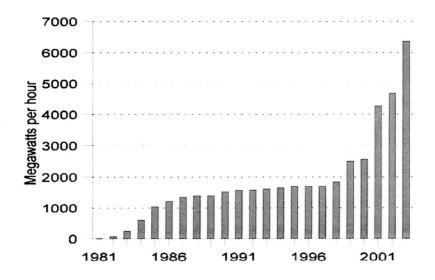

Figure 5. U.S. Installed Wind Energy Capacity, 1981-2003

In the United States, a wind turbine with a generating capacity of 1 MW, placed on a tower situated on a farm, ranch, or other rural land, can generate enough electricity in a year — between 2.4 to 3 million kilowatt-hours (kWh) — to serve the needs of 240 to 300 average U.S. households.[55] However, on average, wind power turbines typically operate the equivalent of less than 40% of the peak (full load) hours in the year due to the intermittency of the wind. Wind turbines are "on-line" —actually generating electricity — only when wind speeds are sufficiently strong (i.e., at least 9 to 10 miles per hour).

Table 6. Installed Wind Energy Capacity by State, August 2004

State	Megawatts	Share
California	2,042.6	32.0%
Texas	1,293.0	20.3%
Minnesota	562.7	8.8%
Iowa	471.2	7.4%
Wyoming	284.6	4.5%
Oregon	259.4	4.1%
Washington	243.8	3.8%
Colorado	223.2	3.5%
New Mexico	206.6	3.2%
Oklahoma	176.3	2.8%
Others	610.5	9.6%
U.S. Total	6,373.9	100.0%

Source: AWEA.

What Is Behind the Rapid Growth of Installed Capacity? Over the past 20 years, the cost of wind power has fallen approximately 90%, while rising natural gas prices have pushed up costs for gas-fired power plants, helping to improve wind energy's market competitiveness.[56] In addition, wind-generated electricity production and use is supported by several federal and state financial and tax incentives, loan and grant programs, and renewable portfolio standards. As of October 2004, renewable portfolio standards have been implemented by 17 states and require that utilities must derive a certain percentage of their overall electric generation from renewable energy sources such as wind power.[57] Environmental and energy security concerns also have encouraged interest in clean, renewable energy sources such as wind power. Finally, rural incomes receive a boost from companies installing wind turbines in rural areas. Landowners have typically received annual lease fees that range from $2,000 to $5,000 per turbine per year depending on factors such as the project size, the capacity of the turbines, and the amount of electricity produced.

Economic Efficiency. The per-unit cost of utility-scale wind energy is the sum of the various costs — capital, operations, and maintenance — divided by the annual energy generation. Utility-scale wind power projects — those projects that generate at least 1 MW of electric power annually for sale to a local utility — account for over 90% of wind power generation in the United States.[58] For utility-scale sources of wind power, a number of turbines are usually built close together to form a wind farm.

In contrast with fossil fuel energy, wind power has no fuel costs. Instead, electricity production depends on the kinetic energy of wind (replenished through atmospheric processes). As a result, its operating costs are lower than costs for power generated from fossil fuels. However, the initial capital investment in equipment needed to set up a utility-scale wind energy system is substantially greater than for competing fossil fuels. Major infrastructure costs include the tower (30 meters or higher) and the turbine blades (generally constructed of fiberglass; up to 20 meters in length; and weighing several thousand pounds). Capital costs generally run about $1 million per megawatt of capacity, so a wind energy system of 10 1.5-megawatt turbines would cost about $15 million. Farmers generally find leasing their land for wind power projects easier than owning projects. Leasing is easier because energy companies can better address the costs, technical issues, tax advantages, and risks of wind projects. Less than 1% of wind power capacity installed nationwide is owned by farmers.[59]

While the financing costs of a wind energy project dominate its competitiveness in the energy marketplace, there are several other factors that also contribute to the economics of utility-scale wind energy production. These include:[60]

- the wind speed and frequency at the turbine location — the energy that can be tapped from the wind is proportional to the cube of the wind speed, so a slight increase in wind speed results in a large increase in electricity generation;
- improvements in turbine design and configuration — the taller the turbine and the larger the area swept by the blades, the more productive the turbine;
- economies of scale — larger systems operate more economically than smaller systems by spreading operations/maintenance costs over more kilowatt-hours;

- transmission and market access conditions (see paragraph below); and
- environmental and other policy constraints — for example, stricter environmental regulations placed on fossil fuel emissions enhance wind energy's economic competitiveness; or, alternately, greater protection of birds or bats,[61] especially threatened or endangered species, could reduce wind energy's economic competitiveness.

Government Support. In addition to market factors, the rate of wind energy system development for electricity generation has been highly dependent on federal government support, particularly a production tax credit that provides a 1.8¢ credit for each kilowatt-hour of electricity produced by qualifying turbines built by the end of 2005 for a 10-year period.[62] In some cases the tax credit may be combined with a five-year accelerated depreciation schedule for wind turbines, as well as with grants, loans, and loan guarantees offered under several different programs.[63] A modern wind turbine can produce electricity for about 2.5¢ to 4¢ per kilowatt hour (including the government subsidy). This implies that the federal production tax credit amounts to 31% to 41% of the cost of production of wind energy. In contrast to wind-generated electricity costs, modern natural-gas-fired power plants produce a kilowatt-hour of electricity for about 5.5¢ (including both fuel and capital costs) when natural gas prices are at $6 per million Btu's.[64]

Natural gas prices have shown considerable volatility since the late 1990s (see **Figure 6**); however, market conditions suggest that the sharp price rise that has occurred since 2002 is unlikely to weaken anytime soon.[65 If natural gas prices continue to be substantially higher than average levels in the 1990s, wind power is likely to be competitive in parts of the country where good wind resources and transmission access can be coupled with the federal production tax credit.

Long-Run Supply Issues. Despite the advantages listed above, U.S. wind potential remains largely untapped, particularly in many of the states with the greatest wind potential, such as North Dakota and South Dakota (see **Figure 7**). Factors inhibiting growth in these states include lack of either (1) major population centers with large electric power demand needed to justify large investments in wind power, or (2) adequate transmission capacity to carry electricity produced from wind in sparsely populated rural areas to distant cities.

Source: DOE, EIA; monthly average industrial prices.

Figure 6. Natural Gas Price, 1998 to 2004

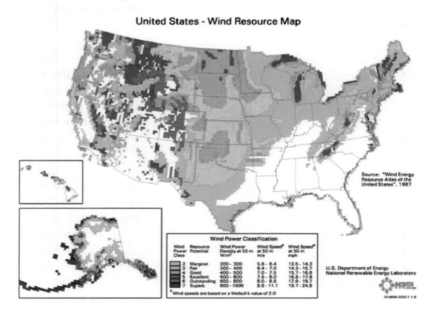

Figure 7. U.S. Areas with Highest Wind Potential

Areas considered most favorable for wind power have average annual wind speeds of about 16 miles per hour or more. The DOE map of U.S. wind potential confirms that the most favorable areas tend to be located in sparsely populated regions, which may disfavor wind-generated electricity production for several reasons. First, transmission lines may be either inaccessible or of insufficient capacity to move surplus wind-generated electricity to distant demand sources. Second, transmission pricing mechanisms may disfavor moving electricity across long distances due to distance-based charges or according to the number of utility territories crossed. Third, high infrastructure costs for the initial hook-up to the power grid may discourage entry, although larger wind farms can benefit from economies of scale on the initial hook-up. Fourth, new entrants may see their access to the transmission power grid limited in favor of traditional customers during periods of heavy congestion. Finally, wind plant operators are often penalized for deviations in electricity delivery to a transmission line that result from the variability in available wind speed.

Environmental Concerns. Three potential environmental issues — impacts on the visual landscape, bird and bat deaths, and noise issues — vary in importance based on local conditions. In some rural localities, the merits of wind energy appear to have split the environmental movement. For example, in the Kansas Flint Hills, local chapters of the Audubon Society and Nature Conservancy oppose installation of wind turbines, saying that they would befoul the landscape and harm wildlife; while Kansas Sierra Club leaders argue that exploiting wind power would help to reduce America's dependence on fossil fuels.

PUBLIC LAWS THAT SUPPORT ENERGY PRODUCTION AND USE BY AGRICULTURE

Clean Air Act Amendments of 1990 (CAAA; P.L. 101-549)

The Reformulated Gasoline and Oxygenated Fuels programs of the CAAA have provided substantial stimuli to the use of ethanol.[66] In addition, the CAAA requires the Environmental Protection Agency (EPA) to identify and regulate air emissions from all significant sources, including on- and off-road vehicles, urban buses, marine engines, stationary equipment, recreational vehicles, and small engines used for lawn and garden equipment. All of these sources are candidates for biofuel use.

Energy Policy Act of 1992 (EPACT; P.L. 102-486)

Energy security provisions of EPACT favor expanded production of renewable fuels. EPACT's alternative-fuel motor fleet program implemented by DOE requires federal, state, and alternative fuel providers to increase purchases of alternative-fueled vehicles. Under this program, DOE has designated neat (100%) biodiesel as an environmentally positive or "clean" alternative fuel.[67] Biodiesel is increasingly being adopted by major fleets nationwide. The U.S. Postal Service, the U.S. military, and many state governments are directing their bus and truck fleets to incorporate biodiesel fuels as part of their fuel base. Currently over 400 fleets use the fuel.[68]

The American Jobs Creation Act of 2004 (AJCA; P.L. 108-357)[69]

The AJCA contains two provisions (Section 301 and 701) that provide tax exemptions for three agri-based renewable fuels: ethanol, biodiesel, and wind energy.

Federal Fuel Tax Exemption for Ethanol (Section 301). This provision provides for an extension and replaces the previous federal ethanol tax incentive (26 U.S.C. 40). The tax credit is revised to allow for blenders of gasohol to receive a federal tax exemption of $0.51 per gallon for every gallon of pure ethanol. Under the revision, the blending level is no longer relevant to the calculation of the tax credit. Instead, the total volume of ethanol used is the basis for calculating the tax.

Federal Fuel Tax Exemption for Biodiesel (Section 301). This provision provides for the first ever federal biodiesel tax incentive — a federal excise tax credit of $1.00 for every gallon of agri-biodiesel (i.e., virgin vegetable oil and animal fat) that is used in blending with petroleum diesel; and a 50¢ credit for every gallon of non-agri-biodiesel (i.e., recycled oils such as yellow grease).

Federal Production Tax Exemption for Wind Energy Systems (Section 710). This provision renews a federal production tax credit that expired on December 31, 2003. The renewed tax credit provides a 1.8¢ credit for a 10-

year period for each kilowatt-hour of electricity produced by qualifying turbines that are built by the end of 2005.

Energy Provisions in the 2002 Farm Bill (P.L. 107-171)[70]

In the 2002 farm bill, Title IX: Energy, Title II: Conservation, and Title VI: Rural Development each contain programs that encourage the research, production, and use of renewable fuels such as ethanol, biodiesel, and wind energy systems.

Federal Procurement of Biobased Products (Title IX, Section 9002). Federal agencies are required to purchase biobased products under certain conditions. A voluntary biobased labeling program is included. Legislation provides funding of $1 million annually through the USDA's Commodity Credit Corporation (CCC) for FY2002-FY2007 for testing biobased products. USDA published final rules in the *Federal Register* (vol. 70, no. 1, pp. 41-50, Jan. 3, 2005). The regulations define what a biobased product is under the statue, identify biobased product categories, and specify the criteria for qualifying those products for preferred procurement.

Biorefinery Development Grants (Title IX, Section 9003). Federal grants are provided to ethanol and biodiesel producers who construct or expand their production capacity. Funding for this program was authorized in the 2002 farm bill, but no funding was appropriated. Through FY2005, no funding had yet been proposed; therefore, no implementation regulations have been developed.

Biodiesel Fuel Education Program (Title IX, Section 9004). Competitively awarded grants are made to nonprofit organizations that educate governmental and private entities operating vehicle fleets, and educate the public about the benefits of biodiesel fuel use. Legislation provides funding of $1 million annually through the CCC for FY2003-FY2007 to fund the program. Final implementation rules were published in the *Federal Register* (vol. 68, no. 189, Sept. 30, 2003).

Energy Audit and Renewable Energy Development Program (Title IX, Section 9005). This program is intended to assist producers in identifying their on-farm potential for energy efficiency and renewable energy use. Funding for this program was authorized in the 2002 farm bill, but through FY2005 no funding has been appropriated. As a result, no implementation regulations have been developed.

Renewable Energy Systems and Energy Efficiency Improvements (Renewable Energy Program) (Title IX; Section 9006). Administered by USDA's Rural Development Agency, this program authorizes loans, loan guarantees, and grants to farmers, ranchers, and rural small businesses to purchase renewable energy systems and make energy efficiency improvements. Grant funds may be used to pay up to 25% of the project costs. Combined grants and loans or loan guarantees may fund up to 50% of the project cost. Eligible projects include those that derive energy from wind, solar, biomass, or geothermal sources. Projects using energy from those sources to produce hydrogen from biomass or water are also eligible. Legislation provides that $23 million will be available annually through the CCC for FY2003-FY2007 for this program. Unspent money lapses at the end of each year.

Prior to each fiscal year, USDA publishes a Notice of Funds Availability (NOFA) in the *Federal Register* inviting applications for the Renewable Energy Program, most recently on May 5, 2004 (*Federal Register*, vol. 69, no. 87). Not all grant applications are accepted. In FY2004, $21 million in grants were offered (compared with $21.7 million in FY2003) for renewable energy projects, including $9.5 million for 37 anaerobic digester projects, $7.9 million for 38 wind power projects, $3.1 million for 13 biomass projects, and $467,167 for 6 geothermal, hybrid, and solar projects.

In order to formalize the program guidelines for receiving and reviewing future loan and loan guarantee applications, USDA published proposed rules in the *Federal Register* (vol. 69, no. 192, Oct. 5, 2004) for a 60-day comment period. Final rules are pending. As a result, no loans or loan guarantees have been offered under this program. USDA estimates that loans and loan guarantees would be more effective than grants in assisting renewable energy projects, because program funds would be needed only for the credit subsidy costs (i.e., government payments made minus loan repayments to the government). USDA estimated that offering a combination of grants, loans, and loan guarantees could equate to as much as $200 million in annual program support.[71]

Hydrogen and Fuel Cell Technologies (Title IX, Section 9007). Legislation requires that USDA and DOE cooperate on research into farm and rural applications for hydrogen fuel and fuel cell technologies. No new budget authority is provided. A draft memorandum of understanding between the two departments has been prepared and is in review.

Biomass Research and Development (Title IX; Section 9008).[72] This provision extends an existing program — created under the Biomass

Research and Development Act (BRDA) of 2000 — that provides competitive funding for research and development projects on biofuels and bio-based chemicals and products, administered jointly by the Secretaries of Agriculture and Energy . Under the BRDA, $49 million per year was authorized for FY2002-FY2005. Section 9008 extends the budget authority through FY2007, but with new funding levels of $5 million in FY2002 and $14 million for FY2003-FY2007 — unspent funds may be carried forward, making the funding total $75 million for FY2002-FY2007. An additional $49 million annually in discretionary funding is also provided for FY2002-FY2007. In FY2004, USDA and DOE awarded a combined total of $25 million in research funding to 21 biomass projects, up from the $23 million awarded in FY2003. The USDA share of this has remained at $14 million for FY2003, FY2004, and FY2005.

Cooperative Research and Development — Carbon Sequestration (Title IX; Section 9009). This provision amends the Agricultural Risk Protection Act of 2000 (P.L. 106-224, Sec. 211) to extend through FY2011 the one-time authorization of $15 million of the Carbon Cycle Research Program, which provides grants to land-grant universities for carbon cycle research with on-farm applications.

Bioenergy Program (Title IX; Section 9010). This is an existing program (7 C.F.R. 1424) in which the Secretary makes payments from the CCC to eligible bioenergy producers — ethanol and biodiesel — based on any year-to-year increase in the quantity of bioenergy that they produce (fiscal year basis). The goal is to encourage greater purchases of eligible commodities used in the production of bioenergy (e.g., corn for ethanol or soybean oil for biodiesel). The 2002 farm bill extended the program through FY2006 and expanded its funding by providing that $150 million be available annually through the CCC for FY2003-FY2006. The final rule for the Bioenergy Program was published in the *Federal Register* (vol. 68, no. 88, May 7, 2003). The FY2003 appropriations act limited spending for the Bioenergy Program funding for FY2003 to 77% ($115.5 million) of the $150 million. In FY2004, no limitations were imposed; however, a $50 million reduction from the $150 million is contained in the FY2005 appropriations act.

Renewable Energy on Conservation Reserve Program (CRP) Lands (Title II; Section 2101). This provision amends Section 3832 of the Farm Security Act of 1985 (1985 farm bill) to allow the use of CRP lands for wind energy generation and biomass harvesting for energy production.

Loans and Loan Guarantees for Renewable Energy Systems (Title VI; Section 6013). This provision amends Section 310B of the Consolidated

Farm and Rural Development Act (CFRDA) (7 U.S.C. 1932(a)(3)) to allow loans for wind energy systems and anaerobic digesters.

Business and Industry Direct and Guaranteed Loans (Title VI; Section 6017(g)(A)). This provision amends Section 310B of CFRDA (7 U.S.C. 1932) to include farmer and rancher equity ownership in wind power projects. Limits range from $25 million to $40 million per project.

Value-Added Agricultural Product Market Development Grants (Title VI; Section 6401(a)(2)). This provision amends Section 231 of CFRDA (7 U.S.C. 1621 note; P.L.106-224) to include farm- or ranch-based renewable energy. Competitive grants are available to assist producers with feasibility studies, business plans, marketing strategies, and start-up capital. Maximum grant amount is $500,000 per project.

Agriculture-Related Provisions in 108th Congress Omnibus Energy Legislation[73]

Major energy legislation (H.R. 6, H.Rept. 108-375, S. 2095) died at the adjournment of the 108th Congress, after stalling in late 2003 and 2004, primarily over high cost and a dispute related to a liability protection provision for MTBE (ethanol's principal oxygenate competitor). Major non-tax provisions in the conference measure and S. 2095 included:[74]

- **Renewable Fuels Standard (RFS)** — Both versions of the energy legislation included a national RFS requiring that 3.1 billion gallons of renewable fuel be used in 2005, increasing to 5.0 billion gallons by 2012.
- **Energy Efficiency Standards** — New statutory efficiency standards would have been established for several consumer and commercial products and appliances. For certain other products and appliances, DOE would have been empowered to set new standards. For motor vehicles, funding would have been authorized for the National Highway Traffic Safety Administration (NHTSA) to set Corporate Average Fuel Economy (CAFE) levels as provided in current law.
- **Energy Production on Federal Lands** — To encourage production on federal lands, royalty reductions would have been provided for marginal oil and gas wells on public lands and the outer continental shelf. Provisions were also included to increase access to federal lands by energy projects — such

as drilling activities, electric transmission lines, and gas pipelines.

Similar provisions may be considered in energy legislation in the 109th Congress.

State Laws and Programs[75]

Several state laws and programs influence the economics of renewable energy production and use by providing incentives for research, production, and consumption of renewable fuels such as biofuels and wind energy systems. In addition, demand for agriculture-based renewable energy is being driven, in part, by state Renewable Portfolio Standards (RPS) that require utilities to obtain set percentages of their electricity from renewable sources by certain target dates. The amounts and deadlines vary, but 17 states now have laws instituting RPS's, with New York being the latest addition.

REFERENCES

[1]   For more information on energy use by the agricultural sector, see CRS Report RL32677, *Energy Use in Agriculture: Background and Issues.*

[2]   Excluding the costs of externalities associated with burning fossil fuels such as air pollution, environmental degradation, and illness and disease linked to emissions.

[3]   For examples of proponent policy positions, see the National Corn Growers Association (NCGA) at [http://www.ncga.com/ethanol/main/index.htm], and the American Soybean Association (ASA) at [http://www.soygrowers.com/policy/].

[4]   Several studies have analyzed the positive gains to commodity prices, farm incomes, and rural employment attributable to increased government support for biofuel production. For examples, see the "For More Information" section at the end of this report.

[5]   Advocates of this position include free-market proponents such as the Cato Institute, and federal budget watchdog groups such as Citizens Against Government Waste and Taxpayers for Common Sense.

[6]   For example, see the Natural Resources Defense Council's fact sheet
        on biomass energy at [http://www.nrdc.org/air/energy/fbiom.asp]. See
        also Robert Wisner and Phillip Baumel, "Ethanol, Exports, and
        Livestock: Will There be Enough Corn to Supply Future Needs?"
        Feedstuffs, no. 30, vol. 76, July 26, 2004.

[7]   See section on "Public Laws That Support Energy Production and Use
        by Agriculture," below, for a listing of major laws supporting farm-
        based renewable energy production.

[8]   For more information on these and other alternative fuels, see CRS
        Report RL30758, *Alternative Transportation Fuels and Vehicles:
        Energy, Environment, and Development Issues*. See also DOE,
        National Renewable Energy Laboratory (NREL), *Introduction to
        Biofuels*, available at [http://www.nrel.gov/clean_energy/biofuels.
        html].

[9]   For more information, see CRS Report RL30369, *Fuel Ethanol:
        Background and Public Policy Issues*.

[10]   For more information, see USDA, Office of Energy Policy and New
        Uses, *The Energy Balance of Corn Ethanol: An Update*, AER-813, by
        Hosein Shapouri, James A. Duffield, and Michael Wang, July 2002
        (hereafter referred to as Shapouri (2002).

[11]   See American Coalition for Ethanol, *Ethanol Production*, at
        [http://www.ethanol.org/ production.html].

[12]   Data sources — corn use for ethanol: USDA, World Agricultural
        Outlook Board, *World Agricultural Supply and Demand Estimates*,
        Dec. 10, 2004; ethanol production: American Coalition for Ethanol,
        Ethanol Production, at [http://www.ethanol.org/production.html].

[13]   Based on a conversion rate of 1.73 GEG per bushel of corn (2.66
        gallons of ethanol per bushel of corn and 0.65 GEG per gallon of
        ethanol).

[14]   U.S. gasoline use: DOE, IEA, *Alternatives to Traditional
        Transportation Fuels 2003*, at [http://www.eia.doe.gov/cneaf/
        alternate/page/datatables/atf1-13_03.html].

[15]   DOE, Energy Efficiency and Renewable Energy (EERE), *Alternative
        Fuel Price Report*, June 29, 2004, at [http://www.eere.energy.gov/
        afdc/resources/pricereport/price_report.html].

[16]   For more information, see section on "Public Laws That Support
        Energy Production and Use by Agriculture," later in this report.

[17]   According to USDA, dry milling is more energy efficient than wet
        milling, particularly when corn co-products are considered; Shapouri
        (2002), p. 5.

[18] H. Shapouri, J. Duffield, and M. Wang, *New Estimates of the Energy Balance of Corn Ethanol*, presented at 2004 Corn Utilization & Technology Conference of the Corn Refiners Association, June 7-9, 2004, Indianapolis, IN; hereafter referred to as Shapouri (2004).

[19] Minnesota Dept. of Agr., *Energy Balance/Life Cycle Inventory for Ethanol, Biodiesel and Petroleum Fuels*, at [http://www.mda.state.mn.us/ethanol/balance.html].

[20] Professor David Pimentel, Cornell University, College of Agriculture and Life Sciences, has researched and published extensive criticisms of corn-based ethanol production. For example, see [http://www.news. cornell.edu/Chronicle/01/8.23.01/Pimentel-ethaol.html].

[21] DOE, EIA, *Annual Energy Outlook 2004 with Projections to 2025.*

[22] Based on USDA's Nov. 12, 2004, WASDE, and using comparable conversion rates.

[23] Assuming yields of 150 bushel per acre.

[24] USDA, ERS, *Feed Situation and Outlook Yearbook*, FDS-2003, April 2003.

[25] For a discussion of potential feed market effects due to growing ethanol production, see Bob Kohlmeyer, "The Other Side of Ethanol's Bonanza," *Ag Perspectives* (World Perspectives, Inc.), Dec. 14, 2004; and R. Wisner and P. Baumel, "Ethanol, Exports, and Livestock: Will There be Enough Corn to Supply Future Needs?" *Feedstuffs*, no. 30, vol. 76, July 26, 2004.

[26] Shapouri (2004), p. 4.

[27] CRS calculations based on DOE energy usage rates.

[28] DOE, EIA, "Outlook for Biomass Ethanol Production and Demand," by Joseph DiPardo, July 30, 2002, available at [http://www.eia. doe.gov/ oiaf/analysispaper/biomass.html]; hereafter referred to as DiPardo (2002).

[29] CRS Report RL30758, *Alternative Transportation Fuels and Vehicles: Energy, Environment, and Development Issues*, p. 24.

[30] For more information on biomass from non-traditional crops as a renewable energy, see the American Bioenergy Association at [http://www.biomass.org/].

[31] Christopher J. Chipello,"Iogen's Milestone: It's Selling Ethanol Made of Farm Waste," *Wall Street Journal*, April 21, 2004.

[32] DOE, EERE, *Biomass Program*, "Cellulase Enzyme Research," available at [http://www. eere.energy.gov/biomass/cellulase_enzyme. html].

[33] For more information, see [http://www.bioproducts-bioenergy.gov].

[34] Argonne National Laboratory, Center for Transportation Research, *Effects of Fuel Ethanol Use on Fuel-Cycle Energy and Greenhouse Gas*, ANL/ESD-38, by M. Wang, C. Saricks, and D. Santini, January 1999, as referenced in DOE, DiPardo (2002).

[35] United Nations, Food and Agricultural Organization (FAO), FAOSTATS.

[36] USDA, Office of Energy Policy and New Uses (OEPNU), *The Economic Impacts of Bioenergy Crop Production on U.S. Agriculture*, AER 816, by Daniel De La Torre Ugarte et al., Feb. 2003; available at [http://www.usda.gov/oce/oepnu/].

[37] Appropriate Technology Transfer for Rural Areas (ATTRA), *Anaerobic Digestion of Animal Wastes: Factors to Consider*, by John Balsam, Oct. 2002, at [http://www.attra.ncat.org]; or Iowa State University, Agricultural Marketing Resource Center, *Anaerobic Digesters/Methane*, at [http://www. agmrc.org/biomass/anaerobic main.html].

[38] U.S. Environmental Protection Agency (EPA), Office of Air and Radiation (OAR), *Managing Manure with Biogas Recovery Systems*, EPA-430-F-02-004,Winter 2002.

[39] USDA, News Release No. 0386.04, Sept. 15, 2004; *Veneman Announces $22.8 Million to Support Renewable Energy Initiatives in 26 States*, available at [http://www.usda.gov/newsroom/0386.04.html].

[40] DOE, EERE, *Methane (Biogas) from Anaerobic Digesters*, at [http://www.eere.energy. gov/consumerinfo/factsheets/ab5.html].

[41] EPA, OAR, *Managing Manure with Biogas Recovery Systems*, EPA-430-F-02-004,Winter 2002.

[42] ATTRA, *Anaerobic Digestion of Animal Wastes: Factors to Consider*, Oct. 2002.

[43] Biodiesel consumption estimates are from DOE, IEA, "Alternatives to Traditional Transportation Fuels 2003, Estimated Data."

[44] National Biodiesel Board (NBB), "U.S. Biodiesel Production Capacity," October 2004, available at [http://www.biodiesel.org/pdf_files/Production%20Capacity_2004.pdf].

[45] Ibid.

[46] DOE, EERE, *Alternative Fuel Price Report*, June 29, 2004, available at [http://www. eere.energy.gov/afdc/resources/pricereport/price_report.html].

[47] DOE, National Renewable Energy Laboratory (NREL), *An Overview of Biodiesel and Petroleum Diesel Life Cycles*, NREL/TP-580-24772,

by John Sheehan et al., May 1998, available at [http://www.afdc.doe.gov/pdfs/3812.pdf].

[48] For more information, see NBB, "Ground-Breaking Biodiesel Tax Incentive Passes," at [http://www.biodiesel.org/resources/press releases/gen/20041011_ FSC_ Passes_Senate.pdf]; see also section on "Public Laws That Support Energy Production and Use by Agriculture" later in this report.

[49] ERS reported the U.S. own-price elasticity for "oils & fats" at -0.027 — i.e., a 10% increase in price would result in a 0.27% decline in consumption. In other words, demand declines only negligibly relative to a price rise. Such inelastic demand is associated with sharp price spikes in periods of supply shortfall. USDA, ERS, *International Evidence on Food Consumption Patterns*, Tech. Bulletin No. 1904, Sept. 2003, p. 67.

[50] For a parallel discussion of feed market consequences from domestic ethanol industry expansion, see Wisner and Baumel in *Feedstuffs*, no. 30, vol. 76, July 26, 2004.

[51] Assuming natural gas is the processing fuel, natural gas demand would increase due to two factors: (1) to produce the steam and process heat in oilseed crushing and (2) to produce methanol used in the conversion step. NREL, *An Overview of Biodiesel and Petroleum Diesel Life Cycles*, NREL/TP-580-24772, by John Sheehan et al., May 1998, p. 19.

[52] For more information on energy consumption by U.S. agriculture, see CRS Report RL32677, *Energy Use in Agriculture: Background and Issues*.

[53] A watt is the basic unit used to measure electric power. A kilowatt (kW) equals 1,000 watts and a megawatt (MW) equals 1,000 kW or 1 million watts. Electricity production and consumption are measured in kilowatt-hours (kWh), while generating capacity is measured in kilowatts or megawatts. If a power plant that has 1 MW of capacity operates nonstop during all 8,760 hours in the year, it will produce 8,760,000 kWh.

[54] American Wind Energy Association (AWEA), at [http://www.awea. org].

[55] An average U.S. household consumes roughly 10,000 kWh per year. Government Accountability Office (GAO), *Renewable Energy: Wind Power's Contribution to Electric Power Generation and Impact on Farms and Rural Communities*, GAO-04-756, Sept. 2004; hereafter referred to as GAO, *Wind Power*, GAO-04-756, Sept. 2004.

[56]   AWEA, *The Economics of Wind Energy*, March 2002.

[57]   Rebecca Smith,"Not Just Tilting Anymore," *Wall Street Journal*, Oct. 14, 2004.

[58]   GAO, *Wind Power*, GAO-04-756, Sept. 2004, p. 66.

[59]   Ibid., p. 6.

[60]   AWEA, *The Economics of Wind Energy*; at [http://www.awea.org].

[61]   Justin Blum, "Researchers Alarmed by Bat Deaths From Wind Turbines," *Washington Post*, by January 1, 2005.

[62]   The federal production tax credit was renewed Oct. 22, 2004 (P.L. 108-357; Sec. 710).

[63]   A five-year depreciation schedule is allowed for renewable energy systems under the Economic Recovery Tax Act of 1981, as amended (P.L. 97-34; Stat. 230, codified as 26 U.S.C. § 168(e)(3)(B)(vi)).

[64]   Rebecca Smith, "Not Just Tilting Anymore," *Wall Street Journal*, Oct. 14, 2004.

[65]   For a discussion of natural gas market price factors, see CRS Report RL32677, *Energy Use in Agriculture: Background and Issues*.

[66]   CRS Report RL30369, *Fuel Ethanol: Background and Public Policy Issues*, p. 6.

[67]   NBB,   "Biodiesel   Emissions,"   at   [http://www.biodiesel.org/pdf_files/emissions.pdf].

[68]   NBB, "Biodiesel 2004 Backgrounder," at [http://www.biodiesel.org/pdf_files/backgrounder.PDF].

[69]   [P.L. 108-357 was signed into law by the President on Oct. 22, 2004. For a discussion of the tax provisions in the bill, as well as information on federal tax credits for other forms of renewable energy, see CRS Report IB10054, *Energy Tax Policy*. For more information on federal production tax credits for biofuels, see CRS Report RL20758, *Alternative Transportation Fuels and Vehicles: Energy, Environment, and Development Issues.*

[70]   USDA, *2002 Farm Bill*, "Title IX — Energy," online information available at [http://www.usda.gov/farmbill/energy_fb.html]. For more information, see CRS Report RL31271, *Energy Provisions of the Farm Bill: Comparison of the New Law with Previous Law and House and Senate Bills.*

[71]   GAO, *Wind Power*, GAO-04-756, Sept. 2004, p. 54-55.

[72]   For more information, see the joint USDA-DOE website at [http://www.bioproducts-bioenergy.gov/].

[73]   For additional related bill contents and more information on non-tax provisions in the bills, see CRS Issue Brief IB10116, *Energy Policy:*

The Continuing Debate and Omnibus Energy Legislation. For a discussion of the tax provisions in the bills, see CRS Issue Brief IB10054, *Energy Tax Policy.*

[74]   For more information, see CRS Report RL32204, *Omnibus Energy Legislation: Comparison of Non-Tax Provisions in the H.R. 6 Conference Report and S. 2095*; and CRS Report RL32078, *Omnibus Energy Legislation: Comparison of Major Provisions in House-and Senate-Passed Versions of H.R. 6, Plus S. 14.*

[75]   For more information on state and federal programs, see *State and Federal Incentives and Laws*, at the DOE's Alternative Fuels Data Center, [http://www.eere.energy.gov/afdc/ laws/incen_laws.html].

In: Biofuels in the Energy Supply System   ISBN: 1-59454-756-4
Editor: Victor I. Welborne, pp. 39-41   © 2006 Nova Science Publishers, Inc.

Chapter 2

BIOFUELS-AT-A-GLANCE [*]

United States Department of Energy

Ethanol and biodiesel are renewable, domestically produced automotive fuels.

The use of biofuels reduces toxic air emissions, greenhouse-gas buildup, dependence on foreign oil, and trade deficits, while supporting local agriculture and rural economies.

Henry Ford designed his first Model T automobile in 1908 to run on ethanol. Rudolf Diesel designed his prototype diesel engine nearly a century ago, to run on peanut oil.

The United States is second to Brazil (where sugar cane is used to make ethanol) in production and use of fuel ethanol and second to Europe (where rapeseed [canola] oil is used to make biodiesel) in the production and use of biodiesel.

ETHANOL FROM GRAIN

Ethanol is used primarily as an octane-boosting, pollution-cutting additive (usually 8% or 10%) to gasoline.

Ethanol's primary additive markets are (1) carbon monoxide non-attainment areas—a market it dominates, and (2) severe ground-level-ozone

---

[*] Extracted from http://www.eere.energy.gov/states/alternatives/

non-attainment areas—a market methyl tertiary butyl ether (MTBE) dominates.

Because of extensive groundwater contamination by MTBE, many states are moving to prohibit MTBE use (already 13 in July 2001). Ethanol, a biodegradable and essentially nontoxic oxygenate, can directly substitute for MTBE. The industry expects to expand sufficiently to meet increased demand from MTBE replacement.

Since 1979, the federal government has provided an excise tax exemption of 5.3 cents per gallon of 10% blend, equal to 53 cents per gallon of ethanol.

Sixteen states provide additional incentives, chiefly state excise tax exemptions or producer credits (as of July 2001: AK, CT, HI, ID, IL, IA, KS, MN, MO, MT, NE, ND, OK, SD, WI, WY).

Federal and state subsidies allow ethanol to compete in the fuel additive market.

E85 (85% ethanol, 15% gasoline) should be used only in flexible-fuel vehicles with modified oxygen sensors and fuel system valves. Flexible-fuel vehicles can also run on gasoline.

Daimler-Chrysler, Ford, and General Motors each offer several models of flex-fuel vehicles with no price premium over other models.

The primary markets for flex-fuel vehicles are federal and state fleets complying with EPAct alternative fuel requirements and individuals wanting to use ethanol.

As of July 2001, 129 retailers in 18 states (AZ, CO, ID, IL, IN, IA, KS, KY, MI, MN, MO, MT, NE, NM, ND, SD, VA, WI) offered E85.

2000 U.S. production of 1.6 billion gallons of ethanol (2.0 billion gallons capacity) used about 600 million bushels of field corn (nation's largest crop, primarily used for animal feed)—about 7% of U.S. corn production.

The sale of corn for ethanol production is credited with increasing corn prices by 25 to 30 cents per bushel (total price about $2 per bushel), greatly benefiting the farm economy, but minimally affecting food prices.

Higher corn prices reduce federal price support payments. This, together with employment taxes at ethanol plants, more than offsets the federal excise tax exemption for ethanol, saving substantial taxpayer money.

Most new ethanol plants are dry-mill plants, relatively small plants, generally located in rural areas, providing valuable jobs and economic development. Since 1990, farmer-owned cooperatives are responsible for 50% of new production capacity.

Ethanol production can stimulate economic growth in small communities throughout the United States. According to a Midwestern Governors' Conference report, ethanol production in the U.S. boosts total employment by 195,200 jobs, improves the U.S. trade balance by $2 billion, and increases net farm income by $4.5 billion.

In: Biofuels in the Energy Supply System   ISBN: 1-59454-756-4
Editor: Victor I. Welborne, pp. 43-44   © 2006 Nova Science Publishers, Inc.

Chapter 3

BIOFUELS IN USE[*]

United States Department of Energy

ETHANOL FROM GRAIN

A project spearheaded by the American Lung Association of Minnesota and the U.S. Department of Energy is bringing E85 (a blend of 85% ethanol, and 15% gasoline) to Minnesota. Since 1998, Minnesota has served as a national pilot market for E85 fuel and flexible fuel vehicles. The project has resulted in the largest E85 fueling network in the United States with 53 fueling stations offering the ethanol-based fuel. The number of E85 flexible fuel vehicles (FFVs) on Minnesota roads has also grown rapidly, and there are currently about 50,000 E85 capable cars, light trucks, and vans on the roadways.

As a major ethanol producer (there are currently 15 ethanol plants in the state), Minnesota is the perfect place to try out the new E85 fuel. By using E85, Minnesota residents have the chance to reduce tailpipe pollution by 25%, reduce greenhouse gas emissions by almost 40%, and support local farmers and ethanol producers. The Minnesota E85 Team works with several local organizations including the Twin Cities Clean Cities Coalition, the Minnesota Timberwolves basketball team, and various fuel providers and FFV manufacturers.

The members of the Minnesota E85 team are: American Lung Association of Minnesota, U.S. Department of Energy, Minnesota Corn

---

[*] Extracted from http://www.eere.energy.gov/states/alternatives/

Growers Association, Minnesota Coalition for Ethanol, Minnesota Department of Commerce, National Ethanol Vehicle Coalition, Minnesota Department of Agriculture, and Ford Motor Company.

ADVANCED BIOETHANOL TECHNOLOGY

Masada Resource Group, based in Birmingham, Alabama, is developing a $130-million waste disposal and recycling facility in Middletown, New York. This facility will recycle plastics, glass, metal, and wastepaper. The plant will also use technology developed in partnership with Tennessee Valley Authority and DOE to convert the remaining cellulosic refuse into 8 million gallons of ethanol each year. The facility is planned for startup in the year 2000. This is the first proposed biomass ethanol plant to use municipal solid waste as a feedstock.

Masada has negotiated contracts with the surrounding municipalities to accept their municipal solid waste, and the municipalities will share in the profits. This waste, which would otherwise be burned or disposed in a landfill, will be used to domestically produce a clean-burning renewable fuel that will displace almost 200,000 barrels of imported oil annually.

BIODIESEL

Some fleet operators were the first to make the switch, fueling their urban bus fleets with biodiesel. Others making the switch include transit bus fleets, heavy-duty truck fleets, airport shuttles, marine and national park boats and vehicles, and military and mining operations.

The Green Team, a San Jose recycling and garbage company, is modifying 95 of its garbage trucks to operate on pure biodiesel (B100). The general manager of this company, Ken Etherington, said that using the new fuel would cut 50,000 pounds of air pollution each year. The Green Team is the first company in the country to use biodiesel fleet-side and in 100% of their vehicles.

In: Biofuels in the Energy Supply System   ISBN: 1-59454-756-4
Editor: Victor I. Welborne, pp. 45-47   © 2006 Nova Science Publishers, Inc.

Chapter 4

ADVANCED BIOETHANOL TECHNOLOGY[*]

United States Department of Energy

Current ethanol production is based on corn grain or other starch or sugar sources which make up only a very small portion of plant material. With advanced bioethanol technology, ethanol can also be made from cellulose and hemicellulose (the components that give plants their structure), which make up the bulk of plant material.

Potential feedstocks for advanced bioethanol technology include corn stover (stalks and husks) and other agricultural residues, wood chips and other forestry residues, paper and other municipal wastes, food processing and other plant-derived industrial wastes, and dedicated energy crops of fast-growing trees or grasses.

Advanced technology bioethanol would supplement rather than replace grain ethanol, but the huge volume of inexpensive available feedstocks offers potential to greatly expand ethanol production and its economic and environmental benefits.

The U.S. Department of Energy National Biofuels Program is supporting research and development to lower the cost of advanced bioethanol technology, so as to make it a marketplace reality, and has set a goal to have commercial demonstration plants using agricultural residues in operation by 2005.

---

[*] Extracted from http://www.eere.energy.gov/states/alternatives/

BIODIESEL

As with ethanol, biodiesel is primarily used as a pollution-reducing additive to conventional diesel, usually in a 20% blend—B20.

U.S. biodiesel production is roughly equally split between soybean oil and recycled restaurant cooking grease—giving biodiesel a reputation for a pleasant "french fry" smell.

2001 biodiesel production was predominantly from soybeans because of a USDA program supporting commodity purchases for increase biofuels production. Both soybean oil and recycled grease are in surplus and biodiesel production uses only a small fraction, so there are no resource constraints.

One primary market for biodiesel is state and federal fleets complying with alternative fuel requirements (EPAct does not cover heavy vehicles, but B20 use—which unlike other alternative fuels requires no new vehicle purchase—gains credits against required light-duty vehicle purchase requirements).

A second primary market is fleets such as city or school bus fleets for which biodiesel makes a visible statement of concern for air quality and customer health.

Biodiesel popularity is growing rapidly with U.S. sales growing from 7 million gallons in 2000 to more than 20 million gallons in 2001. The industry expects to be able to expand capacity rapidly to meet increased demand.

Biodiesel use emits only about half as much carbon monoxide, hydrocarbons, and particulates as petroleum diesel. The cancer-risk contribution of diesel is cut by about 90%. Benefits from B20 use are approximately proportional.

B20 typically costs 8 to 20 cents more than regular diesel.

ETHANOL

Ethanol, also known as ethyl alcohol or grain alcohol, can be used either as an alternative fuel or as an octane-boosting, pollution-reducing additive to gasoline. The U.S. ethanol industry produced more than 3.4 billion gallons in 2004, up from 2.8 billion gallons in 2003 and 2.1 billion gallons in 2002. (Renewable Fuels Association and Renewable Fuels Association Ethanol Industry Outlook 2005). Although this number is small when compared with fossil fuel consumption for transportation, as individual states continue to ban the use of MTBE (Methyl Tertiary Butyl Ether) and with the possibility of a Federal ban, ethanol consumption is due for a significant boost. Because of the increased demand on ethanol as a gasoline additive, efforts to increase supplies are necessary in order to meet the increase in demand. As of the start of 2005, 81 ethanol plants in 20 states have the capacity to produce nearly 4.4 billion gallons annually and an additional 16 plants are under construction to add another 750 million gallons of capacity (RFA).

There are four basic steps in converting biomass to bioethanol:

1. Producing biomass results in the fixing of atmospheric carbon dioxide into organic carbon.
2. Converting this biomass to a useable fermentation feedstock (typically some form of sugar) can be achieved using a variety of different process technologies. These processes for fermentation feedstock production constitute the critical differences among all of the bioethanol technology options.
3. Fermenting the biomass intermediates using biocatalysts (microorganisms including yeast and bacteria) to produce ethanol in a relatively dilute aqueous solution is probably the oldest form of biotechnology developed by humankind.
4. Processing the fermentation product yields fuel-grade ethanol and byproducts that can be used to produce other fuels, chemicals, heat and/or electricity.

Corn and other starches and sugars are only a small fraction of biomass that can be used to make ethanol. Advanced Bioethanol Technology allows fuel ethanol to be made from cellulosic (plant fiber) biomass, such as agricultural forestry residues, industrial waste, material in municipal solid waste, trees, and grasses. Cellulose and hemicellulose, the two main components of plants-and the ones that give plants their structure-are also made of sugars, but those sugars are tied together in long chains. Advanced bioethanol technology can break those chains down into their component sugars and then ferment them to make ethanol. This technology turns ordinary low-value plant materials such as corn stalks, sawdust, or waste paper into fuel ethanol. Not quite lead into gold, but maybe more valuable for the U.S. economy, for cutting air pollution, and for reducing dependence on foreign oil.

In: Biofuels in the Energy Supply System   ISBN: 1-59454-756-4
Editor: Victor I. Welborne, pp. 49-51   © 2006 Nova Science Publishers, Inc.

Chapter 5

BIOFUELS - FREQUENTLY ASKED QUESTIONS*

United States Department of Energy

Question: Will Biofuels Production Affect the Nation's Food Supply?

Answer: The great bulk of U.S. ethanol is made from field corn, the United States' largest agricultural crop (sweet corn for direct human consumption is a minor crop) and one that typically has enough surplus to require price supports. Most field corn is used for animal feed, a lesser amount for food processing. The 7% currently used for ethanol production has little or no impact on the nation's food supply. Ethanol production is credited with modestly increasing corn prices—which is of course welcome by the agricultural community and saves the government money by reducing price support payments—but has minimal impact on consumer food prices.

U.S. biodiesel production is based mostly upon soybean oil and recycled restaurant cooking oil. Currently both are available in surplus and biodiesel production uses a minor amount of each. Although soybean oil is used for cooking oil and various food products, it is also used in a wide range of industrial products. Its use for biodiesel has little impact on food supply.

---

* Extracted from http://www.eere.energy.gov/states/alternatives/

Question: What are Ethanol and Biodiesel Made From?

Answer: Corn is the major feedstock for U.S. ethanol production, but current technology can easily use any starch or sugar source. Brazil's ethanol industry is based on sugar cane. U.S. ethanol producers now also use milo, wheat starch, potato waste, cheese whey, and beverage waste.

Advanced bioethanol technology allows ethanol to be made from virtually any plant or plant-derived material. The cellulose and hemicellulose of the fibrous portion-and great bulk-of plant material is broken down into component sugars and then fermented. As this technology becomes commercialized, it will likely progress from using concentrated "opportunity" feedstocks such as food processing or paper mill wastes, to agricultural and forestry residues such as corn stover and wood chips, to dedicated energy crops such as switchgrass or fast-growing trees.

Biodiesel can be made from any animal fat or vegetable oil. U.S. production capacity is, in fact, about half based on soybean oil and half on recycled cooking oil from restaurants. The European biodiesel industry uses primarily rapeseed (canola) oil.

Question: Where can I find Ethanol or Biodiesel?

Answer: As of 2001, approximately one out of every eight gallons of gasoline sold in the United States contained 8% to 10% ethanol. In some cases, fuel blenders use ethanol to boost octane, but in most cases they use it as an oxygenate to reduce pollution. If your region has high levels of carbon monoxide (16 regions in 10 states), your gasoline probably contains ethanol. If your region has severe levels of ground-level ozone (25 regions in 17 states), your gasoline may contain ethanol, but more likely contains methyl tertiary butyl ether (MTBE). Because, however, of concerns for groundwater contamination, many states are moving to ban MTBE (13 already, as of July 2001), so much of the MTBE use may switch to ethanol over the coming years. Regardless of air pollution requirements, fuel blenders are free to use ethanol as an additive—many do so in the Corn Belt—and you are free to request it or patronize retailers that do use it.

As of July 2001, 129 retail gasoline stations in 18 states (AZ, CO, ID, IL, IN, IA, KS, KY, MI, MN, MO, MT, NE, NM, ND, SD, VA, WI) offered E85 (85% ethanol) for flexible-fuel vehicles. The Alternative Fuels Data Center Alternative Fuel Station Locator or Alternative Fuels Hotline (800-423-1363) can help you locate any stations in your area.

Biodiesel is used mostly by fleet operators, but as of July 2001 was already offered by retail service stations in at least seven states (AZ, CA, HI, ME, MN, NV, SC). The National Biodiesel Board, a trade association for biodiesel producers (573-635-3893), can help you locate these stations or biodiesel producers that market to fleets in your area. Their members include most of the current dedicated producers [7 in 2001 in HI, IA (2), KY, FL, NV, IL].

Question: Are Ethanol Production Facilities Dominated by Large Companies?

Answer: Of U.S. corn ethanol production, about half is in wet-mill plants and half is dry-mill plants. Wet mills are typically large operations that make ethanol along with a slate of food industry products such as corn oil, corn syrup, and animal feed. These plants (7 in 2001) are mostly owned by large companies. Dry-mill plants are typically smaller facilities that produce ethanol as a primary product with a high-protein animal feed known as distillers dried grains, as a co-product. Most dry mills (51 in 2001) are small, local operations that provide markets for local area farmers and jobs and economic development for rural communities. Most are owned by relatively small companies, with many of them owned cooperatively by the local-area farmers providing the corn. Growth in the ethanol industry is predominantly in the dry-mill sector.

Question: Do biofuels work well in conventional vehicles?

Answer: Ethanol in a 10% blend works fine in conventional engines. It increases octane to prevent knocking and reduces carbon monoxide and other toxic air emissions. Daimler-Chrysler, Ford, and General Motors all now offer several models of flexible-fuel vehicles with minor modifications (redesigned oxygen sensors and different seals in the fuel system) standard to accommodate the use of E85 (85% ethanol with 15% gasoline). These models are all made available at the same cost as gasoline-only models.

B20 (20% biodiesel, 80% petroleum diesel) works fine in all diesel engines without modification, reducing toxic air emissions and lubricating to reduce engine wear. Straight biodiesel (B100) requires special management in cold climates. Also, rubber seals, gaskets, and hoses made before 1994 should be replaced when using B100.

Question: Is the United States the Leader in Biofuels Use?

Answer: The United States is second to Brazil, which produces ethanol from the sugar in sugar cane, in both production and use of fuel ethanol. In both production and use of biodiesel, the United States is second to Europe, which makes biodiesel from rapeseed oil.

In: Biofuels in the Energy Supply System   ISBN: 1-59454-756-4
Editor: Victor I. Welborne, pp. 53-61   © 2006 Nova Science Publishers, Inc.

Chapter 6

TECHNOLOGY OVERVIEW[*]

United States Department of Energy

BIOFUELS FOR SUSTAINABLE TRANSPORTATION

Biofuels are liquid transportation fuels made from plant matter instead of petroleum. Ethanol and biodiesel—the primary biofuels today—can be blended with or directly substitute for gasoline and diesel, respectively. Use of biofuels reduces air toxics emissions, greenhouse gas buildup, dependence on imported oil, and trade deficits, while supporting agriculture and rural economies.

Unlike gasoline and diesel, biofuels contain oxygen. Therefore, adding biofuels to petroleum products causes the fuel to combust more completely and reduces air pollution. Also, when fossil fuels such as petroleum are burned, they release carbon dioxide that was captured by plants billions of years ago. This release contributes to the buildup of greenhouse gases that threatens global warming. Carbon dioxide release from burning biofuels, however, is balanced by the carbon dioxide capture by the recent growth of the plant material they are made from. Depending on how much fossil energy is used to grow and process the biomass feedstock, this results in substantially reduced net greenhouse gas emissions.

---

[*] Extracted from http://www.eere.energy.gov/states/alternatives/

- Ethanol, also known as ethyl alcohol or grain alcohol, can be used either as an alternative fuel or an octane-boosting, pollution-reducing additive to gasoline.
- Ethanol from Grain (chiefly the starch in kernels of field corn) is the primary means of current ethanol production in the United States.
- Advanced Bioethanol Technology allows fuel ethanol to also be made from cellulosic (plant fiber) biomass, such as agricultural forestry residues, industrial waste, material in municipal solid waste, and trees and grasses.
- Biodiesel, made from animal fat or vegetable oil, is a renewable pollution-reducing alternative to petroleum diesel.
- E-Diesel, a fuel that uses additives to allow blending of ethanol with diesel, is being developed by several companies.
- Methanol, also known as "wood alcohol," can be made thermochemically from biomass, but is now usually made from natural gas or coal. Research on biomass methanol has waned, because making ethanol from cellulosic material now shows greater promise.

Ethanol from Grain

Biofuels such as ethanol made from corn and biodiesel made from soybeans help support American agriculture. Advanced bioethanol technology will also provide a market for dedicated energy crops such as fast-growing trees for ethanol production.

Ethanol, also known as ethyl alcohol or grain alcohol, is the same clear, colorless, flammable liquid that has been produced as a beverage from time immemorial. Ethanol can be used either as an alternative fuel (usually with 15% gasoline—E85—requires slight engine modification) or as an octane-boosting, pollution-reducing additive to gasoline (usually 8-10%, 5% in California). One out of every eight gallons of gasoline sold in the United States now contains about 10% ethanol. Most of that ethanol is made by fermenting the starch in kernels of field corn, by an industrial version of beverage production. In the future, ethanol could be used as a fuel to power fuel cells.

In Brazil, the world's largest producer of ethanol, sugar made from sugar cane is the primary feedstock. In the United States, the second largest producer and user of fuel ethanol, the primary feedstock is the starch in kernels of field corn (otherwise predominantly used as animal feed). Other current feedstocks include milo, wheat starch, potato waste, cheese whey, and beverage waste, but any starch or sugar source can be used. The Renewable Fuels Association, a trade association for ethanol producers is a good source of additional information.

ADVANCED BIOETHANOL TECHNOLOGY

These pilot-plant-size fermentation tanks produce fuel ethanol with advanced bioethanol technology developed for converting plant materials other than starch or sugar. This research facility at the National Renewable Energy Laboratory tests various processes that produce ethanol from a wide range of biomass feedstocks. (Warren Gretz)

Corn and other starches and sugars are only a small fraction of "biomass," the wide range of plant and plant-derived waste materials. Cellulose and hemicellulose, the two main components of plants—and the ones that give plants their structure—are also made of sugars, but those sugars are tied together in long chains. Advanced bioethanol technology can break those chains down into their component sugars and then ferment them to ethanol. U.S. Department of Energy (DOE) Biofuels Program scientists and engineers are at the forefront of bioethanol technology research. This technology turns ordinary low-value plant materials such as corn stalks, sawdust, or waste paper into fuel ethanol. Not quite lead into gold, but maybe more valuable for the U.S. economy, for cutting air pollution, and for reducing dependence on foreign oil. To help improve this technology and ready it for commercial operation, the DOE researchers and their industrial partners use the DOE Bioethanol Pilot Plant, a fully integrated biomass-to-ethanol production facility that can turn as much as one ton per day of corn stalks or other plant material into transportation fuel.

Biodiesel

Biodiesel use dramatically reduces toxic emissions without new vehicle purchase, so is a great way for operators of fleets such as city and school buses to show their concern for their customers' health.

Biodiesel, made by transforming animal fat or vegetable oil with alcohol, can be directly substituted for diesel either as neat fuel (B100) or as an oxygenate additive (typically 20%—B20). B20 earns credits for alternative fuel use under the Energy Policy Act of 1992, the only fuel that does not require purchase of new vehicles. In Europe, the largest producer and user of biodiesel, the fuel is usually made from rapeseed (canola) oil. In the United States, the second largest producer and user of biodiesel, the fuel is usually made from soybean oil or recycled restaurant grease. The National Biodiesel Board, a trade association for biodiesel producers is a good source of additional information.

Biofuels Resource

Ethanol from Grain

Current U.S. ethanol production is based primarily upon field corn kernels. Approximately 90% of field corn is fed directly to livestock and only 7% (600 million bushels) is used to make ethanol (1.6 billion gallons production and 2 billion gallons capacity) in 2000. Therefore, the general supply of corn is not likely to limit ethanol production in the near future. A Governors' Ethanol Coalition report,"Ability of the U.S. Ethanol Industry to Replace MTBE," found that the industry could double production capacity within two years— to 3.5 billion gallons per year, by 2004.

Advanced Bioethanol Technology

Using advanced bioethanol technology to produce fuel ethanol from cellulose and hemicellulose opens up a vast resource to supplement grain ethanol production. Forest residues, urban wastes, agricultural residues, mill residues, and energy crops have been estimated to be able to supply more than 500 mill dry tons of biomass, enough to make more than 50 billion gallons of ethanol (equivalent to 33 billion gallons of gasoline) or about one quarter the current U.S. consumption of gasoline. Corn stover (corn stalks and husks) and wheat straw alone could be made into 10-15 billion gallons of ethanol.

Biodiesel

Biodiesel popularity is growing rapidly, with U.S. sales increasing from about 7 million gallons in 2000 to more than 20 million gallons in 2001. Both soybean oil and recycled restaurant cooking oil—from which biodiesel is made—are in surplus and biodiesel production uses only a small portion of each, so there is no resource constraint. (Because of incentives from a U.S. Department of Agriculture program supporting commodity purchases for increased biofuel production, 2001 biodiesel production was predominantly from soybean oil, but recycled cooking oil will likely again roughly balance soybean oil, once the effect of that program is over.) 2001 production was near capacity for dedicated biodiesel producers, but most are expanding and the detergent and fatty acid industries could provide another 30 to 50 million gallons of capacity, if needed to meet demand.

How Does It Work?

Ethanol from Grain

Modern fuel ethanol technology is highly sophisticated and efficient, but the basic process is similar to that of making alcoholic beverages. Most fuel ethanol is biologically produced by using an enzyme to convert starch to sugar and a yeast to ferment sugar to ethanol.

Of U.S. corn ethanol production, about half is in wet-mill plants and half in dry-mill plants. The former are typically large operations that produce ethanol along with a slate of food industry products such as corn oil, corn syrup, and corn sugar. The latter are typically smaller facilities that produce ethanol as a primary product with a high-protein animal feed known as distillers dried grains as a co-product.

Used as an additive, ethanol requires no engine modification. Because—unlike gasoline—it contains oxygen, fuel with ethanol added burns more completely—particularly in older or high-mileage vehicles—reducing carbon monoxide and other toxic emissions. Ethanol additive also boosts octane to avoid engine knocking (about 3 octane points for 10% blend). Use as an alternative fuel—typically with 15% gasoline to aid cold-weather starting (E85)—requires minor modifications including redesigned oxygen sensors and different seals in the fuel system. E85 flex-fuel vehicles qualify as alternative fuel vehicles and Daimler-Chrysler, Ford, and General Motors all offer several models that can use E85 at the same cost as gasoline-only models. Because ethanol contains about one-third less energy per gallon than gasoline, E85 vehicles will get fewer miles per gallon.

Advanced Bioethanol Technology

Starches and sugars constitute only a small portion of plant matter. The bulk of most plants—and waste materials derived from plant matter—

consists of cellulose, hemicellulose, and lignin. Although the fibrous cellulose and hemicellulose are not digestible by humans or the common yeast now used for ethanol production, they are carbohydrates, consisting of sugars in polymer chains. The challenge is to break those chains down into their component sugars for fermentation. Also, the sugars from hemicellulose are different, so cannot be fermented by the standard yeast. Because of the sophisticated conversion technology required, making ethanol from cellulosic biomass is currently more expensive than making it from corn grain, but because the feedstocks would be inexpensive the expectation is for equal or lower costs in the long run. The U.S. Department of Energy Biomass Program is a good source of information.

Biodiesel

Esters are compounds of alcohol and organic acids. Fatty acid methyl ester, commonly known as biodiesel, is made by bonding methanol to animal fat or vegetable oil. The process is relatively straightforward, but must consistently achieve prescribed standards to minimize the risk of damaging expensive diesel engines.

Because it is oxygenated, biodiesel dramatically reduces air toxins, carbon monoxide, soot, small particles, and hydrocarbon emissions by 50% or more, reducing the cancer-risk contribution of diesel up to 90% with pure biodiesel. Air quality benefits are roughly proportional for diesel/biodiesel mixtures. Biodiesel's superior lubricity helps reduce engine wear, even as a small percentage additive.

The most common use of biodiesel—20% biodiesel, 80% diesel—requires no engine modifications. Because it gels at higher temperatures compared to petroleum diesel, however, pure biodiesel requires special management in cold climates. Also, rubber seals, gaskets, and hoses made before 1994 should be replaced when using straight biodiesel. Engine manufacturers' warranties vary with regard to biodiesel use, so engine manufacturers should be contacted. Biodiesel contains slightly less energy than petroleum diesel but it's more dense, so fuel economy tends to fall 7% for every 10% biodiesel in your fuel blend.

Growth in the ethanol industry can benefit smaller and agricultural communities.

Two basic markets exist for biofuels: the fuel additive market and the alternative fuel market. The fuel additive market exists for a number of reasons, including reducing carbon monoxide emissions, reducing chemical emissions that lead to smog formation, and as a substitute for methyl tertiary butyl ether (MTBE).

Ethanol from Grain

The Clean Air Act Amendments of 1990 mandated the sale of oxygenated fuels in regions with unhealthy levels of carbon monoxide, a poisonous gas that interferes with the ability of blood to transport oxygen. The use of ethanol can help offset carbon monoxide emissions in areas with unhealthy levels of CO. Most regions use blends of 8-10% ethanol in their gasoline to comply.

For smog/ozone, the Clean Air Act Amendments also direct regions severely failing to meet standards for ground-level ozone—a strong oxidant that causes respiratory problems and the main pollutant we usually associate with "smog"— to use "reformulated gasoline" (RFG). The EPA required reformulated gasoline to contain an oxygenate as a specification of these standards. Although ethanol is widely used in the Midwest, fuel manufacturers have predominantly used methyl tertiary butyl ether (MTBE), a petroleum-derived oxygenate, up to this point and approximately 25% of gasoline today contains MTBE as an additive. As a result of surface and groundwater contamination, however, California and 12 other states have moved to phase out the use of MTBE. Nontoxic and biodegradable, ethanol can directly and easily replace MTBE use and could be added to all U.S. gasoline. The replacement of MTBE as an oxygenate will substantially increase demand for ethanol as an additive.

Currently, the ethanol industry is responsible for approximately 200,000 jobs. In 2000, 56 plants produced a total of more than 1.6 billion gallons of ethanol, still less than 2% of total automotive fuel use but enough to be an additive for one in eight gallons of gasoline. From 1996 to 2002, the ethanol industry will add $51 billion to the U.S. economy.

Ethanol production creates domestic jobs in plant construction, plant operation, plant maintenance, and plant support in the surrounding local communities where it is produced. Growth in the ethanol industry offers potential for overall economic development and additional employment in smaller and agricultural communities. The U.S. Department of Agriculture (USDA) estimates that a 100 million gallon ethanol production facility can create 2,250 local jobs for a single community.

About half of ethanol production is from wet mills, which produce a slate of food industry products such as corn syrup, corn sugar, and corn oil, plus gluten feed and gluten meal animal feeds, in addition to ethanol. Growth in wet mill production will therefore be mostly a function of demand for those products, more than for ethanol, so will likely be modest.

The other half of current ethanol production is from dry mills, which while they have ethanol as a primary product, also have distillers dried grains as an economically important co-product. The market for this high-protein animal feed is finite and saturation could significantly affect dry-mill plant economics, slowing growth at some point in the future. In the meantime, most recent and anticipated future expansion of the ethanol industry is coming from the dry mill industry, which is growing rapidly.

Advanced Bioethanol Technology

Advanced bioethanol technology ethanol will supplement rather than replace corn-grain ethanol, but will also greatly expand the potential feedstock supply and make ethanol production an option for all parts of the country. Corn stover (stalks and husks) is a likely key feedstock, but non-cornbelt areas can use biomass such as forestry residues or municipal waste or grow dedicated energy crops. The U.S. Department of Energy Biomass Program is spearheading the effort to improve advanced bioethanol technology and has a goal of having commercial demonstration plants in operation by 2005.

As the cost effectiveness of conversion technology increases, industrial companies are becoming increasingly interested in commercializing the technology to produce ethanol from cellulose and hemicellulose. In fact, within the next few years, the first commercial alternative-feedstock ethanol plants will begin operation in the United States.

Biodiesel

U.S. biodiesel use grew from 7 million gallons in 2000 to more than 20 million gallons in 2001 and continued dramatic growth is expected. Although alternative fuel vehicle requirements of the Energy Policy Act of 1992 do not apply to heavy-duty vehicles, federal, state, and alternative-fuel-provider fleets get credit for biodiesel use against their requirement to purchase alternative-fuel light-duty vehicles—without any new vehicle purchase. An important additional market is for fleets such as city and school buses where biodiesel use visibly demonstrates concern for air quality and customer health. Biodiesel is also starting to see significant non-transportation use such as in boilers and diesel generators.

- 1.0 imperial gallon = 4.55 liter = 1.20 US gallon
- 1.0 liter = 0.264 US gallon = 0.220 imperial gallon
- 1.0 US bushel = 0.0352 m3 = 0.97 UK bushel = 56 lb, 25 kg (corn or sorghum) = 60 lb, 27 kg (wheat or soybeans) = 40 lb, 18 kg (barley)

Areas and Crop Yields

- 1.0 hectare = 10,000 m2 (an area 100 m x 100 m, or 328 x 328 ft) = 2.47 acres
- 1.0 km2 = 100 hectares = 247 acres
- 1.0 acre = 0.405 hectares
- 1.0 US ton/acre = 2.24 t/ha
- 1 metric tonne/hectare = 0.446 ton/acre
- 100 g/m2 = 1.0 tonne/hectare = 892 lb/acre
- for example, a "target" bioenergy crop yield might be: 5.0 US tons/acre (10,000 lb/acre) = 11.2 tonnes/hectare (1120 g/m2)

Biomass Energy

- Cord: a stack of wood comprising 128 cubic feet (3.62 m3); standard dimensions are 4 x 4 x 8 feet, including air space and bark. One cord contains approx. 1.2 U.S. tons (oven-dry) = 2400 pounds = 1089 kg
- 1.0 metric tonne wood = 1.4 cubic meters (solid wood, not stacked)
- Energy content of wood fuel (HHV, bone dry) = 18-22 GJ/t (7,600-9,600 Btu/lb)
- Energy content of wood fuel (air dry, 20% moisture) = about 15 GJ/t (6,400 Btu/lb)
- Energy content of agricultural residues (range due to moisture content) = 10-17 GJ/t (4,300-7,300 Btu/lb)
- Metric tonne charcoal = 30 GJ (= 12,800 Btu/lb) (but usually derived from 6-12 t air-dry wood, i.e. 90-180 GJ original energy content)
- Metric tonne ethanol = 7.94 petroleum barrels = 1262 liters

- ethanol energy content (LHV) = 11,500 Btu/lb = 75,700 Btu/gallon = 26.7 GJ/t = 21.1 MJ/liter. HHV for ethanol = 84,000 Btu/gallon = 89 MJ/gallon = 23.4 MJ/liter
- ethanol density (average) = 0.79 g/ml (= metric tonnes/m3)
- Metric tonne biodiesel = 37.8 GJ (33.3 - 35.7 MJ/liter)
- biodiesel density (average) = 0.88 g/ml (= metric tonnes/m3)

Fossil Fuels

- Barrel of oil equivalent (boe) = approx. 6.1 GJ (5.8 million Btu), equivalent to 1,700 kWh. "Petroleum barrel" is a liquid measure equal to 42 U.S. gallons (35 Imperial gallons or 159 liters); about 7.2 barrels oil are equivalent to one tonne of oil (metric) = 42-45 GJ.
- Gasoline: US gallon = 115,000 Btu = 121 MJ = 32 MJ/liter (LHV). HHV = 125,000 Btu/gallon = 132 MJ/gallon = 35 MJ/liter
- Metric tonne gasoline = 8.53 barrels = 1356 liter = 43.5 GJ/t (LHV); 47.3 GJ/t (HHV)
- gasoline density (average) = 0.73 g/ml (= metric tonnes/m3)
- Petro-diesel = 130,500 Btu/gallon (36.4 MJ/liter or 42.8 GJ/t)
- petro-diesel density (average) = 0.84 g/ml (= metric tonnes/m3)
- Note that the energy content (heating value) of petroleum products per unit mass is fairly constant, but their density differs significantly – hence the energy content of a liter, gallon, etc. varies between gasoline, diesel, kerosene.
- Metric tonne coal = 27-30 GJ (bituminous/anthracite); 15-19 GJ (lignite/sub-bituminous) (the above ranges are equivalent to 11,500-13,000 Btu/lb and 6,500-8,200 Btu/lb).
- Note that the energy content (heating value) per unit mass varies greatly between different "ranks" of coal. "Typical" coal (rank not specified) usually means bituminous coal, the most common fuel for power plants (27 GJ/t).
- Natural gas: HHV = 1027 Btu/ft3 = 38.3 MJ/m3; LHV = 930 Btu/ft3 = 34.6 MJ/m3
- Therm (used for natural gas, methane) = 100,000 Btu (= 105.5 MJ)

Carbon Content of Fossil Fuels and Bioenergy Feedstocks

- coal (average) = 25.4 metric tonnes carbon per terajoule (TJ)
- 1.0 metric tonne coal = 746 kg carbon
- oil (average) = 19.9 metric tonnes carbon / TJ
- 1.0 US gallon gasoline (0.833 Imperial gallon, 3.79 liter) = 2.42 kg carbon
- 1.0 US gallon diesel/fuel oil (0.833 Imperial gallon, 3.79 liter) = 2.77 kg carbon
- natural gas (methane) = 14.4 metric tonnes carbon / TJ
- 1.0 cubic meter natural gas (methane) = 0.49 kg carbon
- carbon content of bioenergy feedstocks: approx. 50% for woody crops or wood waste; approx. 45% for graminaceous (grass) crops or agricultural residues

ENVIRONMENTAL BENEFITS

Tens of millions of Americans live in areas that don't meet at least one federal air quality standard. In 1990, Congress passed the Clean Air Act Amendments to combat high emission levels of carbon monoxide and the creation of ground-level ozone by petroleum-based transportation fuels. This Act specifically required the use of oxygenated fuels during winter months in areas exceeding standards for carbon monoxide and "reformulated gasoline" in areas exceeding standards for ground-level ozone. In order to fulfill the oxygenate requirement, ethanol is now blended with gasoline in nearly all the carbon monoxide non-attainment areas and is added to a modest but growing portion of reformulated gasoline (RFG).

Although diesel fuel regulations don't require the use of oxygenates per se, oxygen-containing renewable diesel alternatives such as biodiesel (fatty-acid methyl ester made from vegetable oil or animal fat) and e-diesel (ethanol blended with diesel) can dramatically reduce emissions from diesel engines. Biodiesel used straight or in a typical 20 percent blend with petroleum diesel reduces visible smoke, odor, and toxic emissions as follows:

Emission	B100	B20
Carbon Monoxide	-47%*	-12%*
Hydrocarbons	-67%*	-20%*
Particulates	-48%*	-12%*
Nitrogen oxides	+10%*	+2%*
Air Toxics	-60%-90%	-12%-20%
Mutagenicity-	80%-90%	-20%

Figure ES-A Environmental Protection Agency Draft Technical Report EPA420-P-02-001 "A Comprehensive Analysis of Biodiesel Impacts on Exhaust Emissions" (PDF 765 KB) Download Acrobat Reader.

Biofuels are essentially nontoxic and biodegrade readily. Every gallon of biofuels used reduces the hazard of toxic petroleum product spills from oil tankers and pipeline leaks (average of 12 million gallons per year, more than what spilled from the Exxon Valdez, according to the U.S. Department of Transportation). In addition, using biofuels reduces the risk of groundwater contamination from underground gasoline storage tanks (more than 46 million gallons per year from 16,000 small oil spills, according to the General Accounting Office), and runoff of vehicle engine oil and fuel.

The U.S. transportation sector is responsible for one-third of our country's carbon dioxide (CO_2) emissions, the principal greenhouse gas contributing to global warming. Combustion of biofuels also releases CO_2, but because biofuels are made from plants that just recently captured that CO_2 from the atmosphere-rather than billions of years ago-that release is largely balanced by CO_2 uptake for the plants' growth. The CO_2 released when biomass is converted into biofuels and burned in truck or automobile engines is recaptured when new biomass is grown to produce more biofuels. Depending upon how much fossil energy is used to grow and process the biomass feedstock, this results in substantially reduced net greenhouse gas emissions. Modern, high-yield corn production is relatively energy intense, but the net greenhouse gas emission reduction from making ethanol from corn grain is still about 20%. Making biodiesel from soybeans reduces net emissions nearly 80%. Producing ethanol from cellulosic material also involves generating electricity by combusting the non-fermentable lignin. The combination of reducing both gasoline use and fossil electrical production can mean a greater than 100% net greenhouse gas emission reduction.

For more information about biofuels and global climate change, read Biofuels-A Solution for Climate Change: Our Changing Earth, Our

Changing Climate (PDF 651 KB) or Bioethanol-The Climate-Cool Fuel (PDF 358 KB) Download Acrobat Reader.

Environmental Benefits of BioPower

Biomass electricity is typically generated through boiler/steam turbine plants, but with three key differences: the fuel is renewable, there is less than 0.1% sulfur (an acid rain ingredient) in biomass fuels, and less air pollutants are produced. More specific environmental benefits for biomass power are:

Reduced Sulfur Dioxide Emissions

Most forms of biomass contain very small amounts of sulfur, therefore a biomass power plant emits very little sulfur dioxide (SO_2), an acid rain precursor. Coal, however, usually contains up to 5% sulfur. Biomass mixed with coal can significantly reduce the power plant's SO_2 emissions compared to a coal-only operation.

Reduced Nitrogen Oxide Emissions

Recent biomass tests at several coal-fired power plants in the U.S. have demonstrated that NOx emissions can be reduced relative to coal-only operations. By carefully adjusting the combustion process, NOx reductions at twice the rate of biomass heat input have been documented.

Reduced Carbon Emissions

Plants absorb CO_2 during their growth cycle when managed in a sustainable cycle, like raising energy crops or replanting harvested areas. Biomass Power generation can be viewed as a way to recycle carbon. Thus, Biomass Power generation can be considered a carbon-neutral power generation option.

Reducing Other Emissions

Landfills produce methane (CH_4) from decomposing biomass materials. Decomposing animal manure, whether it is land-applied or left uncovered in a lagoon also generates methane. Methane, which is the main component of natural gas, is normally discharged directly into the air, but it can be captured and used as a fuel to generate electricity and heat.

Reduced Odors

Using animal manure and landfill gas for energy production can reduce odors associated with conventional disposal or land applications.

Environmental Benefits of Biobased Products

Many of the products now made from petroleum (e.g., petrochemicals) could be made from renewable biomass. The basic molecules in petrochemicals are hydrocarbons. In plant resources, the basic molecules are carbohydrates, proteins, and plant oils. Both plant and petroleum molecules can be processed to create building blocks for industry to manufacture a wide variety of consumer goods, including plastics, solvents, paints, adhesives, and drugs.

During the last century hydrocarbon feedstocks have dominated as industrial inputs. However, reserves of petroleum are finite and, while expected to last well into the next century, could be significantly depleted as the world population grows and standards of living improve in developing countries. Renewable plant resources will be one way to supplement hydrocarbon resources and meet increasing worldwide needs for consumer goods. We are currently witnessing the emergence of new biobased commercial and industrial chemicals, pharmaceuticals, and products. Utilization of these products on a larger scale has the potential to make an impact on reducing U.S. reliance on fossil fuels and sequestering carbon, both of which benefit the environment.

NATIONAL ENERGY SECURITY

Cheap oil fuels America's economy. According to the Energy Information Administration (EIA) , in 2002, the United States consumed 19.656 million barrels of petroleum (crude oil and petroleum products) per day, or about one-quarter of total world oil production. More than half (62%) was imported oil. The EIA projected total petroleum consumption in 2025 at 28.3 million barrels per day - increasing to 70% dependency on foreign imports (EIA Annual Energy Outlook 2004). Most of this demand for oil over the next two decades is in the transportation sector. As sources of domestic oil supplies disappear, the nation's increasing reliance on imported oil makes the United States vulnerable to oil supply disruptions, and threatens America's economic and energy security.

Energy Security and the U.S. Transportation Sector

The transportation sector relies heavily on oil, accounting for two-thirds of U.S. petroleum use in 2002 and this level of consumption is expected to continue through 2025 (EIA Annual Energy Outlook 2004 Figure 102). Throughout this forecast period, the level of gasoline consumption is projected to increase from 8.9 to 13.3 million barrels per day (EIA Annual Energy Outlook 2004, Figure 103). In addition, more than 50% of the fuel used by the transportation sector is imported-far more than any other part of the U.S. economy. This makes transportation particularly vulnerable to the risks of relying on foreign oil.

Reducing the transportation sector's reliance on oil is clearly the key to improving our nation's energy security. Together with measures such as improving vehicle fuel efficiency, using biomass derived ethanol and biodiesel as additives to gasoline and diesel can help offset some of our demand for petroleum. U.S. ethanol production, with corn as the primary feedstock, totaled 2.81 billion gallons in 2003 (up from 2.14 billion gallons in 2002) and production is projected to increase to 3.2 billion gallons in 2025 (EIA Annual Energy Outlook 2004, Figure 104), with about 27 percent of the growth from conversion of cellulosic biomass (such as wood and agricultural residues). Use of biodiesel has also increased significantly - about 20 million gallons of biodiesel were produced in the United States in 2001. (Renewable Fuels Association, and Alternative Fuel News (PDF 625 KB) (Download Acrobat Reader)). Incrementally increasing the biofuels content of motor vehicle fuel (gasoline and diesel) from 1.2 to 4.0 percent between 2002 and 2016 would displace a total of 2.9 billion barrels of crude oil (Urbanchuk, J.M. "An Economic Analysis of Legislation for a Renewable Fuels Requirement for Highway Motor Fuels." AUS Consultants, November 7, 2001 (PDF 130 KB).

U.S. Oil Imports

To keep up with America's ever-increasing demand for oil, the United States has steadily increased its dependence on foreign oil since 1985. In 1993, total imports as a share of petroleum products supplied broke the 50% mark for the first time. Today total imports of 11.5 million barrels per day comprise 58.2% of petroleum products supplied (EIA Monthly Energy Review December 2001, Table 1.8).

The statistics above are based on gross imports and ignore U.S. exports of petroleum. Net imports, which take into account U.S. exports of petroleum, give a better indication of the big picture—the fraction of oil consumed that could not have been supplied by domestic sources. In 2000, net imports totaled 10.4 million barrels per day, or 53% of petroleum products supplied. Net imports are projected to increase to 16.6 million barrels per day, or 62% of petroleum supplied by 2020 (EIA Annual Energy Outlook 2002, Figure 80 and Table 15).

America's heavy reliance on imported oil jeopardizes our nation's energy, economic, and environmental security, particularly in the transportation sector. In the current situation, the United States has little control over oil supply disruptions and oil price fluctuations. The necessity of maintaining a stable supply of imported oil imposes foreign policy constraints, and in times of crisis, forces the U.S. military into action. See U.S. Military and Oil.

Depletion of U.S. Oil Reserves

Declining U.S. oil reserves and falling domestic production from aging oil fields are key factors in America's increasing dependence on foreign imports. In addition, America has already developed the bulk of its known and easily accessible low-cost deposits. The following statistics from EIA clearly summarize the problem:

- U.S. proven oil reserves have declined by an estimated 16.3 billion barrels from 39.0 billion barrels in 1970 to 22.7 billion barrels at the end of 2002 (EIA U.S. Crude Oil, Natural Gas, and Natural Gas Liquids Reserves 2002 Annual Report, Table 6 (PDF 126 KB)). This less than 2% of the world's known oil reserves (EIA International Energy Outlook 2003, Table 11).
- Domestic oil production has been steadily declining since 1970. U.S. petroleum production is expected to decrease slightly from 9.2 million barrels per day in 2002 to 8.6 million barrels per day by 2025, but oil consumption in the United States is expected to rise from 19.6 million barrels per day in 2002 to 28.3 million barrels per day in 2025, a 44% increase (EIA Annual Energy Outlook 2004, Figure 99 (PDF 126 KB)).

The combination of dwindling U.S. oil reserves and increasing oil demand make it impossible for the United States to significantly improve energy security by using more domestic petroleum, even if the United States were to tap every remaining oil deposit in America. That would just delay the inevitable; and the United States would still have to reduce its use of petroleum products and turn to alternative transportation fuel sources such as biofuels in order to gain a secure energy future.

U.S. Vulnerability to Oil Supply Disruptions

While there is no question that the United States is increasingly dependent on foreign oil, the level of oil dependence doesn't really give a full indication of how vulnerable the United States is to an oil supply disruption. If the U.S. oil supply came from many small producers, and one of them suddenly stopped exporting oil, then the impact on oil supply and prices would be small, even at a high level of dependence. However, this is not the case; today, four major producers provide over nearly 70% of the U.S. oil supply: Canada, Mexico, Venezuela, and the Persian Gulf region (Bahrain, Iran, Iraq, Kuwait, Qatar, Saudi Arabia and the United Arab Emirates).

- Canada, Mexico, and Venezuela, combined, supplied over 45% of the oil supplied to the United States in 2002 (EIA's Petroleum Supply Annual 2002 (PDF 22 KB), Volume 1, Table 21). Trade agreements with Canada and Mexico, and the proximity of these sources should make their supplies less vulnerable to disruptions.

- In 2002, the Persian Gulf supplied nearly 20% of U.S. imported petroleum (EIA Annual Energy Review 2002 (PDF 15 KB), Table 5.4). This region will continue to increase its influence in world oil markets, as oil supplies in other regions are exhausted, because over half the world's known oil reserves are concentrated in the Persian Gulf (EIA International Energy Outlook 2003).

The United States first experienced oil supply disruptions from the Persian Gulf region in the 1970s, when two sudden and sharp oil price hikes rocked the American economy. Since then, additional disruptions in oil supply, such as those occurring during the 1979 Iranian revolution and the 1990 invasion of Kuwait by Iraq, reinforce the need to reduce America's dependence on Middle Eastern oil. See Oil Supply Disruptions and the Economy.

The ability of the United States to offset a major oil supply disruption has improved little since the 1970s. Several factors are contributing to America's increasing vulnerability.

- By 2025, oil and oil production facilities will be concentrated in the Asia/Pacific region (EIA Annual Energy Outlook 2003, Figure 101). At these levels, a supply disruption from this one region would have an immediate impact on oil supplies and prices worldwide.
- The U.S. government's emergency supply of crude oil, the strategic petroleum reserve (SPR) , provides less protection from an oil supply disruption than in previous years, because of America's increasing demand for oil. The maximum days of inventory protection peaked at 118 days in 1985; it currently down to 53 days.
- About 61% of the increase in petroleum demand over the next two decades will be met by an increase in production by members of OPEC rather than by non-OPEC suppliers. By 2025, OPEC production is expected to be more than 25 million barrels per day higher than it was in 2001 (EIA International Energy Outlook 2003, Figure 37).
- Although these factors are all indicators of America's vulnerability, there is no real way to estimate the probability of disruption. In the near term, greater diversity of oil import sources can reduce America's vulnerability to oil supply disruptions. In the long term, promoting energy efficiency and producing and using fuels from renewable, domestic biomass resources—particularly in the

transportation sector—will ease our dependence on foreign oil imports and improve our nation's energy security. In addition to its direct displacement of imported oil, biofuels production and use creates the infrastructure to respond to future oil supply disruptions. The greater the percentage of transportation fuel coming from biofuels, the more quickly the industry will be able to increase production if needed to meet an emergency situation.

REFERENCES:

[1]   Ehrman, Tina. Determination of Acid-Soluble Lignin in Biomass. NREL-LAP-004. Golden, CO: National Renewable Energy Laboratory, September 9,1996.

[2]   Voet, Donald; Voet, Judith G. Biochemistry. New York: John Wiley, 1990.

[3]   Domalski, E.S.; Milne T.A., ed. Thermodynamic Data for Biomass Materials and Waste Components. The ASME Research Committee on Industrial and Municipal Wastes, New York: The American Society of Mechanical Engineers, 1987.

[4]   Fengel, Dietrich; Gerd, Wegener. Wood Chemistry, Ultrastructure, and Reactions. Berlin-New York: Walter de Gruyter, 1989.

[5]   Smook, Gary A. Handbook for Pulp and Paper Technologists. Vancouver, BC: Angus Wilde Publications, 1992.

[6]   Ehrman, Tina. Standard Method for Determination of Total Solids in Biomass. NREL-LAP-001. Golden, CO: National Renewable Energy Laboratory, October 28,1994.

[7]   "Biofuels Glossary." (September 1986). Solar Technical Information Program, Solar Energy Research Institute.

[8]   Milne, T.A.; Brennan, A.H.; Glenn, B.H. Sourcebook of Methods of Analysis for Biomass Conversion and Biomass Conversion Processes. SERI/SP-220-3548. Golden, CO: Solar Energy Research Institute, February 1990.

In: Biofuels in the Energy Supply System   ISBN: 1-59454-756-4
Editor: Victor I. Welborne, pp. 77-97   © 2006 Nova Science Publishers, Inc.

Chapter 8

FUEL ETHANOL: BACKGROUND AND PUBLIC POLICY ISSUES[*]

Brent D. Yacobucci and Jasper Womach

SUMMARY

In light of a changing regulatory environment, concern has arisen regarding the future prospects for ethanol as a motor fuel. Ethanol is produced from biomass (mainly corn) and is mixed with gasoline to produce cleaner-burning fuel called "gasohol" or "E10."

The market for fuel ethanol, which consumes 6% of the nation's corn crop, is heavily dependent on federal subsidies and regulations. A major impetus to the use of fuel ethanol has been the exemption that it receives from the motor fuels excise tax. Ethanol is expensive relative to gasoline, but it is subject to a federal tax exemption of 5.4 cents per gallon of gasohol (or 54 cents per gallon of pure ethanol). This exemption brings the cost of pure ethanol, which is about double that of conventional gasoline and other oxygenates, within reach of the cost of competitive substances. In addition, there are other incentives such as a small ethanol producers tax credit. It has been argued that the fuel ethanol industry could scarcely survive without these incentives.

The Clean Air Act requires that ethanol or another oxygenate be mixed with gasoline in areas with excessive carbon monoxide or ozone pollution.

---

[*] Excerpted from CRS Report RL30369 dated Updated March 22, 2000.

The resulting fuels are called oxygenated gasoline (oxyfuel) and reformulated gasoline (RFG), respectively. Using oxygenates, vehicle emissions of volatile organic compounds (VOCs) have been reduced by 17%, and toxic emissions have been reduced by approximately 30%. However, there has been a push to change the oxygenate requirements for two reasons. First, methyl tertiary butyl ether (MTBE), the most common oxygenate, has been found to contaminate groundwater. Second, the characteristics of ethanol-blended RFG-along with high crude oil prices and supply disruptions-led to high Midwest gasoline prices in Summer 2000, especially in Chicago and Milwaukee.

Uncertainties about future oxygenate requirements, as both federal and state governments consider changes, have raised concerns among farm and fuel ethanol industry groups and have prompted renewed congressional interest in the substance. Without the current regulatory requirements and incentives, or something comparable, much of ethanol's market would likely disappear. Expected changes to the reformulated gasoline requirements could either help or hurt the prospects for fuel ethanol (subsequently affecting the corn market), depending on the regulatory and legislative specifics. As a result, significant efforts have been launched by farm interests, the makers of fuel ethanol, agricultural states, and the manufacturers of petroleum products to shape regulatory policy and legislation.

This report provides background concerning various aspects of fuel ethanol, and a discussion of the current related policy issues.

INTRODUCTION

Ethanol (ethyl alcohol) is an alcohol made by fermenting and distilling simple sugars. Ethyl alcohol is in alcoholic beverages and it is denatured (made unfit for human consumption) when used for fuel or industrial purposes.[1] The biggest use of fuel ethanol in the United States is as an additive in gasoline. It serves as an as an oxygenate (to prevent air pollution from carbon monoxide and ozone), as an octane booster (to prevent early ignition, or "engine knock"), and as an extender of gasoline. In purer forms, it can also be used as an alternative to gasoline in automobiles designed for its use. It is produced and consumed mostly in the Midwest, where corn--the main feedstock for ethanol production--is produced.

The initial stimulus to ethanol production in the mid-1970s was the drive to develop alternative and renewable supplies of energy in response to the oil embargoes of 1973 and 1979. Production of fuel ethanol has been encouraged by a partial exemption from the motor fuels excise tax. Another impetus to fuel ethanol production has come from corn producers anxious to expand the market for their crop. More recently the use of fuel ethanol has been stimulated by the Clean Air Act Amendments of 1990, which require oxygenated or reformulated gasoline to reduce emissions of carbon monoxide (CO) and volatile organic compounds (VOCs).

While oxygenates reduce CO and VOC emissions, they also can lead to higher emissions of nitrogen oxides, precursors to ozone formation. While reformulated gasoline has succeeded in reducing ground-level ozone, the overall effect of oxygenates on ozone formation has been questioned. Furthermore, ethanol's main competitor in oxygenated fuels, methyl tertiary butyl ether (MTBE), has been found to contaminate groundwater. This has led to a push to ban MTBE, or eliminate the oxygenate requirements altogether. High summer gasoline prices in the Midwest, especially in Chicago and Milwaukee, where oxygenates are required, have added to the push to remove the oxygenate requirements. The trade-offs between air quality, water quality, and consumer price have sparked congressional debate on these requirements. In addition, there has been a long-running debate over the tax incentives that ethanol-blended fuels receive.

Fuel ethanol is used mainly as a low concentrate blend in gasoline, but can also be used in purer forms as an alternative to gasoline. In 1999, 99.8% of fuel ethanol consumed in the United States was in the form of "gasohol" or "E10" (blends of gasoline with up to 10% ethanol).[2]

Fuel ethanol is produced from the distillation of fermented simple sugars (e.g. glucose) derived primarily from corn, but also from wheat, potatoes and other vegetables, as well as from cellulosic waste such as rice straw and sugar cane (bagasse). The alcohol in fuel ethanol is identical to ethanol used for other purposes, but is treated (denatured) with gasoline to make it unfit for human consumption.

Ethanol and the Agricultural Economy

Corn constitutes about 90% of the feedstock for ethanol production in the United States. The other 10% is largely grain sorghum, along with some barley, wheat, cheese whey and potatoes. Corn is used because it is a relatively low cost source of starch that can be converted to simple sugars, fermented and distilled. It is estimated by the U. S. Department of Agriculture (USDA) that about 615 million bushels of corn will be used to produce about 1.5 billion gallons of fuel ethanol during the 2000/2001 corn marketing year.[3] This is 6.17% of the projected 9.755 billion bushels of corn utilization.[4]

Producers of corn, along with other major crops, receive farm income support and price support. Farms with a history of corn production will receive "production flexibility contract payments" of about $1.186 billion during the 2000/2001 corn marketing year. Emergency economic assistance (P.L. 106-224) more than double the corn contract payments. Corn producers also are guaranteed a minimum national average price of $1.89/bushel under the nonrecourse marketing assistance loan program.[5]

Table 1. Corn Utilization, 2000/2001 Forecast

Quantity(million bushels)	Share of Total Use	
Livestock feed & residual	5,775	59.2%
Food, seed & industrial:	1,980	19.9%
Fuel alcohol	615	6.2%
High fructose corn syrup	550	5.5%
Glucose & dextrose	220	2.2%
Starch	225	2.6%
Cereals & other products	190	1.9%
Beverage alcohol	130	1.3%
Seed	20	0.2%
Exports	2000	20.1%
TOTAL USE	9,775	100.00%
TOTAL PRODUCTION	9,968	

Source: Basic data are from USDA, Economic Research Service, Feed Outlook, March 10, 2000.

The added demand for corn created by fuel ethanol raises the market price for corn above what it would be otherwise. Economists estimate that when supplies are large, the use of an additional 100 million bushels of corn raises the price by about 4¢ per bushel. When supplies are low, the price impact is greater. The ethanol market is particularly welcome now, when the average price received by farmers is forecast by USDA to average about $1.80 per bushel for the 2000/01 marketing year. This price would be the lowest season average since 1986. The ethanol market of 615 million bushels of corn, assuming a price impact of about 25¢ per bushel on all corn sales, means a possible $2.4 billion in additional sales revenue to corn farmers. In the absence of the ethanol market, lower corn prices probably would stimulate increased corn utilization in other markets, but sales revenue would not be as high. The lower prices and sales revenue would be likely to result in higher federal spending on corn payments to farmers, as long as corn prices were below the price triggering federal loan deficiency subsidies.

ETHANOL REFINING AND PRODUCTION

According to the Renewable Fuels Association, about 55% of the corn used for ethanol is processed by "dry" milling plants (a grinding process) and the other 45% is processed by "wet" milling plants (a chemical extraction process). The basic steps of both processes are as follows. First, the corn is processed, with various enzymes added to separate fermentable sugars. Next, yeast is added to the mixture for fermentation to make alcohol. The alcohol is then distilled to fuel-grade ethanol that is 85-95% pure.[6] Finally, for fuel and industrial purposes the ethanol is denatured with a small amount of a displeasing or noxious chemical to make it unfit for human consumption.[7] In the U.S. the denaturant for fuel ethanol is gasoline.

Ethanol is produced largely in the Midwest corn belt, with almost 90% of production occurring in five states: Illinois, Iowa, Nebraska, Minnesota and Indiana. Because it is generally less expensive to produce ethanol close to the feedstock supply, it is not surprising that the top five corn-producing states in the U.S. are also the top five ethanol-producers. Most ethanol use is in the metropolitan centers of the Midwest, where it is produced. When ethanol is used in other regions, shipping costs tend to be high, since ethanol-blended gasoline cannot travel through petroleum pipelines.

This geographic concentration is an obstacle to the use of ethanol on the East and West Coasts. The potential for expanding production geographically is a motivation behind research on ethanol, since if regions could locate production facilities closer to the point of consumption, the costs of using ethanol could be lessened. Furthermore, if regions could produce fuel ethanol from local crops, there would be an increase in regional agricultural income.

Table 2. Top 10 Ethanol Producers by Capacity, 2000

Million Gallons Per Year	
Archer Daniels Midland (ADM)	797
Minnesota Corn Processors	110
Williams Energy Services	100
Cargill	100
New Energy Corp	85
Midwest Grain Products	78
High Plains Corporation	70
Chief Ethanol	62
AGP	52
A.E. Staley	45
Chief Ethanol	40
All Others	508
U.S. Total	2007

Source: Renewable Fuels Association, Ethanol Industry Outlook 2001..

Ethanol production is also concentrated among a few large producers. The top five companies account for approximately 60% of production capacity, and the top ten companies account for approximately 75% of production capacity. (See Table 2.) Critics of the ethanol industry in general--and specifically of the ethanol tax incentives--argue that the tax incentives for ethanol production equate to "corporate welfare" for a few large producers.[8]

Overall, domestic ethanol production capacity is approximately 2.0 billion gallons per year. Consumption is expected to increase from 1.7 billion gallons per year in 2000 to approximately 2.6 billion gallons per year in 2005. Production will need to increase proportionally to meet the increased demand.[9] However, if the Clean Air Act is amended to limit or ban MTBE, ethanol production capacity may expand at a faster rate. This is especially true if MTBE is banned while maintaining the oxygenate requirements, since ethanol is the most likely substitute for MTBE.

Fuel is not the only output of an ethanol facility, however. Co-products play an important role in the profitability of a plant. In addition to the primary ethanol output, the corn wet milling generates corn gluten feed, corn gluten meal, and corn oil, and dry milling creates distillers grains. Corn oil is used as a vegetable oil and is higher priced than soybean oil. Approximately 12 million metric tons of gluten feed, gluten meal, and dried distillers grains are produced in the United States and sold as livestock feed annually. A major market for corn gluten feed and meal is the European Union, which imported nearly 5 million metric tons of gluten feed and meal during FY1998.

Revenue from the ethanol byproducts help offset the cost of corn. The net cost of corn relative to the price of ethanol (the ethanol production margin) and the difference between ethanol and wholesale gasoline prices (the fuel blending margin) are the major determinants of the level of ethanol production. Currently, the ethanol production margin is high because of the low price of corn. At the same time, the wholesale price of gasoline is increasing against the price of ethanol, which encourages the use of ethanol as an octane enhancer.

FUEL CONSUMPTION

Approximately 1.4 billion gallons of ethanol fuel were consumed in the United States in 1999, mainly blended into E10 gasohol. While large, this figure represents only 1.2% of the approximately 125 billion gallons of gasoline consumption in the same year.[10] According to DOE, ethanol consumption is expected to grow to 2.6 billion gallons per year in 2005 and 3.3 billion gallons per year in 2020. This would increase ethanol's market share to approximately 1.5% by 2005. This 1.5% share is projected to remain constant through 2020.[11]

The most significant barrier to wider use of fuel ethanol is its cost. Even with tax incentives for ethanol producers (see the section on Economic Effects), the fuel tends to be more expensive than gasoline per gallon. Furthermore, since fuel ethanol has a somewhat lower energy content, more fuel is required to travel the same distance. This energy loss leads to an approximate 3% decrease in miles-per-gallon vehicle fuel economy with gasohol.[12]

However, ethanol's chemical properties make it very useful for some applications, especially as an additive in gasoline. Major stimuli to the use of ethanol have been the oxygenate requirements of the Reformulated Gasoline (RFG) and Oxygenated Fuels programs of the Clean Air Act.[13] Oxygenates are used to promote more complete combustion of gasoline, which reduces carbon monoxide and volatile organic compound (VOC) emissions.[14] In addition, oxygenates can replace other chemicals in gasoline, such as benzene, a toxic air pollutant (see the section on Air Quality).

The two most common oxygenates are ethanol and methyl tertiary butyl ether (MTBE). MTBE, primarily made from natural gas or petroleum products, is preferred to ethanol in most regions because it is generally much less expensive, is easier to transport and distribute, and is available in greater supply. Because of different distribution systems and blending processes (with gasoline), substituting one oxygenate for another can lead to significant cost increases.

Despite the cost differential, there are several possible advantages of using ethanol over MTBE. Ethanol contains 35% oxygen by weight--twice the oxygen content of MTBE. Furthermore, since ethanol is produced from agricultural products, it has the potential to be a sustainable fuel, while MTBE is produced from natural gas and petroleum, fossil fuels. In addition, ethanol is readily biodegradable, eliminating some of the potential concerns about groundwater contamination that have surrounded MTBE (see the section on MTBE).

Both ethanol and MTBE also can be blended into otherwise non-oxygenated gasoline to raise the octane rating of the fuel, and therefore improve its combustion properties. High-performance engines and older engines often require higher octane fuel to prevent early ignition, or "engine knock." Other chemicals may be used for the same purpose, but some of these alternatives are highly toxic, and some are regulated as pollutants under the Clean Air Act.[15] Furthermore, since these additives do not contain oxygen, they do not result in the same emissions reductions as oxygenated gasoline.

In purer forms, ethanol can also be used as an alternative to gasoline in vehicles specifically designed for its use, although this only represents approximately 0.2% of ethanol consumption in the U.S. The federal government and state governments, along with businesses in the alternative fuel industry, are required to purchase alternative-fueled vehicles by the Energy Policy Act of 1992.[16] In addition, under the Clean Air Act Amendments of 1990, municipal fleets can use alternative fuel vehicles to mitigate air quality problems. Blends of 85% ethanol with 15% gasoline (E85), and 95% ethanol with 5% gasoline (E95) are currently considered alternative fuels by the Department of Energy.[17] The small amount of gasoline added to the alcohol helps prevent corrosion of engine parts, and aids ignition in cold weather.

**Table 3. Estimated U. S. Consumption of Fuel Ethanol,
MTBE and Gasoline (Thousand Gasoline-Equivalent Gallons)**

	1994	1996	1998	2000	(projected)
E85		80	694	1,727	3,283
E95		140	2,699	59[a]	59
Ethanol in Gasohol (E10)		845,900	660,200	916,000	908,700
MTBE in Gasoline		2,108,800	2,749,700	2,915,600	3,111,500
Gasoline[b]		113,144,000	117,783,000	122,849,000	127,568,000

Source: Department of Energy, Alternatives to Traditional Transportation Fuels 1998 .

[a] A major drop in E95 consumption occurred between 1997 and 1998 because of a significant decrease in the number of E95-fueled vehicles in operation (347 to 14), due to the elimination of an ethanol-fueled bus fleet in California.

[b] Gasoline consumption includes ethanol in gasohol and MTBE in gasoline.

Approximately 1.7 million gasoline-equivalent gallons (GEG)[18] of E85, and 59 thousand GEG of E95 were consumed in 1998, mostly in Midwestern states.[19] (See Table 3.) One reason for the relatively low consumption of E85 and E95 is that there are relatively few vehicles on the road that operate on these fuels. In 1998, approximately 13,000 vehicles were fueled by E85 or E95,[20] [21] as compared to approximately 210 million gasoline- and diesel-fueled vehicles that were on the road in the same year.[22] One obstacle to the use of alternative fuel vehicles is that they are generally more expensive than conventional vehicles, although this margin has decreased in recent years with newer technology. Another obstacle is that, as was stated above, fuel ethanol is generally more expensive than gasoline or diesel fuel. In addition, there are very few fueling sites for E85 and E95, especially outside of the Midwest.

RESEARCH AND DEVELOPMENT IN CELLULOSIC FEEDSTOCKS

For ethanol to play a more important role in U.S. fuel consumption, the fuel must become price-competitive with gasoline. Since a major part of the total production cost is the cost of feedstock, reducing feedstock costs could lead to lower wholesale ethanol costs. For this reason, there is a great deal of interest in the use of cellulosic feedstocks, which include low-value waste products, such as recycled paper, or dedicated fuel crops, such as switch grass. A dedicated fuel crop is one that would be grown and harvested solely for the purpose of fuel production.

However, as the name indicates, cellulosic feedstocks are high in cellulose, and cellulose cannot be fermented. Cellulose must first be broken down into simpler carbohydrates, and this can add an expensive step to the process. Therefore, research has focused on both reducing the process costs for cellulosic ethanol, and improving the availability of cellulosic feedstocks.

On August 12, 1999, the Clinton Administration announced the Biobased Products and Bioenergy Initiative, which aims to triple the use of fuels and products derived from biomass by 2010.[23] Research and development covers all forms of biobased products, including lubricants, adhesives, building materials, and biofuels. Because federal research into cellulosic ethanol is ongoing, it is likely that funding would increase under the initiative.

COSTS AND BENEFITS OF FUEL ETHANOL

Economic Effects

Given that a major constraint on the use of ethanol as an alternative fuel, and as an oxygenate, is its high price, ethanol has not been competitive with gasoline as a fuel. Wholesale ethanol prices, before incentives from the federal government and state governments, are generally twice that of wholesale gasoline prices. With federal and state incentives, however, the effective price of ethanol is much lower. Furthermore, gasoline prices have risen recently, making ethanol more attractive.

The primary federal incentive to support the ethanol industry is the 5.4¢ per gallon exemption that blenders of gasohol (E10) receive from the 18.4¢ federal excise tax on motor fuels.[24] Because the exemption applies to blended fuel, of which ethanol comprises only 10%, the exemption provides for an effective subsidy of 54¢ per gallon of pure ethanol. (See Table 4.)

It is argued that the ethanol industry could not survive without the tax exemption. An economic analysis conducted in 1998 by the Food and Agriculture Policy Research Institute, in conjunction with the congressional debate over extension of the tax exemption, concluded that ethanol production from corn would decline from 1.4 billion gallons per year, and stabilize at about 290 million gallons per year, if the exemption were eliminated.[25]

Table 4. Price of Pure Ethanol Relative to Gasoline

July 1998 to June 1999	
Ethanol Wholesale Price[a]	103 ¢/gallon
Alcohol Fuel Tax Incentive	54 ¢/gallon
Effective Price of Ethanol	49 ¢/gallon
Gasoline Wholesale Price[b]	46 ¢/gallon

Source: Hart's Oxy-Fuel News; Energy Information Agency, Petroleum Marketing Monthly.
[a]This is the average price for pure ("neat") ethanol.
[b]This is the average rack price for regular conventional gasoline (i.e. non-oxygenated, standard octane).

The tax exemption for ethanol is criticized by some as a corporate subsidy,[26] because, in this view, it encourages the inefficient use of agricultural and other resources, and deprives the Highway Trust Fund of needed revenues.[27] In 1997, the General Accounting Office estimated that the tax exemption would lead to approximately $10.4 billion in foregone Highway Trust Fund revenue over the 22 years from FY1979 to FY2000.[28] The petroleum industry opposes the incentive because it also results in reduced use of petroleum.

Proponents of the tax incentive argue that ethanol leads to better air quality, and that substantial benefits flow to the agriculture sector due to the increased demand for corn created by ethanol. Furthermore, they argue that the increased market for ethanol leads to a stronger U.S. trade balance, since a smaller U.S. ethanol industry would lead to increased imports of MTBE to meet the demand for oxygenates.[29]

Air Quality

One of the main motivations for ethanol use is improved air quality. Ethanol is primarily used in gasoline to meet minimum oxygenate requirements of two Clean Air Act programs. Reformulated gasoline (RFG)[30] is used to reduce vehicle emissions in areas that are in severe or extreme nonattainment of National Ambient Air Quality Standards (NAAQS) for ground-level ozone.[31] Ten metropolitan areas, including New York, Los Angeles, Chicago, Philadelphia, and Houston, are covered by this requirement, and many other areas with less severe ozone problems have opted into the program, as well. In these areas, RFG is used year-round. By contrast, the Oxygenated Fuels program operates only in the winter months in 20 areas[32] that are listed as carbon monoxide (CO) nonattainment areas.[33]

EPA states that RFG has led to significant improvements in air quality, including a 17% reduction in volatile organic compounds (VOCs) emissions from vehicles, and a 30% reduction in toxic emissions. Furthermore, according to EPA "ambient monitoring data from the first year of the RFG program (1995) also showed strong signs that RFG is working. For example, detection of benzene (one of the air toxics controlled by RFG, and a known human carcinogen) declined dramatically, with a median reduction of 38% from the previous year."[34]

However, the need for oxygenates in RFG has been questioned. Although oxygenates lead to lower emissions of VOCs, and CO, they may lead to higher emissions of nitrogen oxides (NOX). Since all three contribute to the formation of ozone, the National Research Council recently concluded that while RFG certainly leads to improved air quality, the oxygenate requirement in RFG may have little overall impact on ozone formation.[35] Furthermore, the high price of Midwest gasoline in Summer 2000 has raised further questions about the RFG program (see the section on Phase 2 Reformulated Gasoline).

Evidence that the most widely-used oxygenate, methyl tertiary butyl ether (MTBE), contaminates groundwater has led to a push by some to eliminate the oxygen requirement in RFG. MTBE has been identified as an animal carcinogen, and there is concern that it is a possible human carcinogen. In California, MTBE will be banned as of December 31, 2002, and the state is lobbying Congress for a waiver to the oxygen requirement (see section on MTBE). Other states, such as states in the Northeast, are also seeking waivers.

If the oxygenate requirements were eliminated, some refiners claim that the environmental goals of the RFG program could be achieved through cleaner, although potentially more costly, gasoline that does not contain any oxygenates.[36] These claims have added to the push to remove the oxygen requirement and allow refiners to produce RFG in the most cost-effective manner, whether or not that includes the use oxygenates. However, some environmental groups are concerned that an elimination of the oxygenate requirements would compromise air quality gains resulting from the current standards, since oxygenates also displace other harmful chemicals in gasoline. This potential for "backsliding" is a result of the fact that the current performance of RFG is substantially better that the Clean Air Act requires. If the oxygenate standard were eliminated, environmental groups fear that refiners would only meet the requirements of the law, as opposed to maintaining the current overcompliance.

While the potential ozone benefit from oxygenates in RFG has been questioned, there is little dispute that the winter Oxy-Fuels program has led to lower emissions of CO. The Oxy-Fuels program requires oxygenated gasoline in the winter months to control CO pollution in NAAQS nonattainment areas for the CO standard. However, this program is small relative to the RFG program.[37]

The air quality benefits from purer forms of ethanol can also be substantial. Compared to gasoline, use of E85 and E95 can result in a 30-50% reduction in ozone-forming emissions. And while the use of ethanol also leads to increased emissions of acetaldehyde, a toxic air pollutant, as defined by the Clean Air Act, these emissions can be controlled through the use of advanced catalytic converters.[38] However, as was stated above, these purer forms of ethanol have not seen wide use.

Climate Change

Another potential environmental benefit from ethanol is the fact that it is a renewable fuel. Proponents of ethanol argue that over the entire fuel-cycle[39] it has the potential to reduce greenhouse gas emissions from automobiles relative to gasoline, therefore reducing the risk of possible global warming.

Because ethanol (C_2H_5OH) contains carbon, combustion of the fuel necessarily results in emissions of carbon dioxide (CO_2), the primary greenhouse gas. However, since photosynthesis (the process by which plants convert light into chemical energy) requires absorption of CO_2, the growth cycle of the feedstock crop can serve--to some extent--as a "sink" that absorbs some of these emissions. In addition to CO_2 emissions, the emissions of other greenhouse gases may increase or decrease depending on the fuel cycle.[40]

According to Argonne National Laboratory, using E10, vehicle greenhouse gas emissions (measured in grams per mile) are approximately 1% lower than with the same vehicle using gasoline. With improvements in production processes, by 2010, the reduction in greenhouse gas emissions from ethanol relative to gasoline could be as high as 8-10% for E10, while the use of E95 could lead to significantly higher reductions.[41]

While some studies have called into question the efficiency of the ethanol production process, most recent studies find a net energy gain.[42] If true, then the overall reductions in greenhouse gas emissions would be diminished, due to higher fuel consumption during the production process.

Energy Security

Another frequent argument for the use of ethanol as a motor fuel is that it reduces U.S. reliance on oil imports, making the U.S. less vulnerable to a fuel embargo of the sort that occurred in the 1970s, which was the event that initially stimulated development of the ethanol industry. According to Argonne National Laboratory, with current technology the use of E10 leads to a 3% reduction in fossil energy use per vehicle mile, while use of E95 could lead to a 44% reduction in fossil energy use.[43]

However, other studies contradict the Argonne study, suggesting that the amount of energy needed to produce ethanol is roughly equal to the amount of energy obtained from its combustion, which could lead to little or no reductions in fossil energy use.[44] Thus, if the energy used in ethanol production is petroleum-based, ethanol would do nothing to contribute to energy security. Furthermore, as was stated above, fuel ethanol only displaces approximately 1.2% of gasoline consumption in the United States. This small market share led GAO to conclude that the ethanol tax incentive has done little to promote energy security.[45] Furthermore, since ethanol is currently dependent on the U.S. corn supply, any threats to this supply (e.g. drought), or increases in corn prices, would negatively affect the cost and/or supply of ethanol. This happened when high corn prices caused by strong export demand in 1995 contributed to an 18% decline in ethanol production between 1995 and 1996.

POLICY CONCERNS AND CONGRESSIONAL ACTIVITY

Recent congressional interest in ethanol fuels has mainly focused on three sets of issues: 1) implementation of Phase 2 of the RFG program; 2) a possible phase-out of MTBE; and 3) the alcohol fuel tax incentives.

Phase 2 Reformulated Gasoline

Under the new Phase 2 requirements of the RFG program, which took effect in 2000, gasoline sold in the summer months (beginning June 1) must meet a tighter volatility standard.[46] Reid Vapor Pressure (RVP) is a measure of volatility, with higher numbers indicating higher volatility. Because of its physical properties, ethanol has a higher RVP than MTBE. Therefore, to make Phase 2 RFG with ethanol, the gasoline, called RBOB,[47] must have a lower RVP. This low-RVP fuel is more expensive to produce, leading to higher production costs for ethanol-blended RFG.

Before the start of Phase 2, estimates of the increased cost to produce RBOB for ethanol-blended RFG ranged from 2 to 4 cents per gallon, to as much as 5 to 8 cents per gallon.[48] In Summer 2000, RFG prices in Chicago and Milwaukee were considerably higher than RFG prices in other areas, and it has been argued that the higher production cost for RBOB was one cause. However, not all of the price difference is attributable to the new Phase 2 requirements or the use of ethanol. Conventional gasoline prices in the Midwest were also high compared with gasoline prices in other areas. High crude oil prices, low gasoline inventories, pipeline problems, and uncertainties over a patent dispute pushed up prices for all gasoline in the Midwest.

To decrease the potential for price spikes, on March 15, 2001, EPA announced that Chicago and Milwaukee will be allowed to blend slightly higher RVP reformulated gasoline during the summer months.[49] This action is not a change in regulations but a revision of EPA's enforcement guidelines. In addition to EPA's action, one possible regulatory option that has been suggested to control summer RFG prices is a more significant increase in the allowable RVP under Phase 2. Although the volatility standard is set by the Clean Air Act, the Environmental Protection Agency (EPA) is currently reviewing whether credits from ethanol's improved performance on carbon monoxide emissions are possible as an offset to its higher volatility. Legislative options have included eliminating the oxygenate standard for RFG, or suspending the program entirely. However, some in the petroleum industry suggest that additional changes to fuel requirements could further disrupt gasoline supplies. No bills to address the RVP issue have been introduced in the 107th Congress.

MTBE

Another key issue involving ethanol is the current debate over MTBE. Since MTBE, a possible human carcinogen, has been found in groundwater in some states (especially in California), there has been a push both in California and nationally to ban MTBE.[50] In March 1999, California's

Governor Davis issued an Executive Order requiring that MTBE be phased out of gasoline in the state by December 31, 2002. Arizona, Connecticut, Iowa, Minnesota, Nebraska, New York, and South Dakota have also instituted limits or bans on MTBE. In July 1999, an advisory panel to EPA recommended that MTBE use should be "reduced substantially."[51]

A possible ban on MTBE could have serious consequences for fuel markets, especially if the oxygenate requirements remain in place. Since ethanol is the second most used oxygenate, it is likely that it would be used to replace MTBE. However, there is not currently enough U.S. production capacity to meet the potential demand. Therefore, it would likely be necessary to phase out MTBE over time, as opposed to an immediate ban. Furthermore, the consumer price for oxygenated fuels would likely increase because ethanol, unlike MTBE, cannot be shipped through pipelines and must be mixed close to the point of sale, adding to delivery costs. Increased demand for oxygenates could also be met through imports from countries such as Brazil, which is a leader worldwide in fuel ethanol production, and currently has a surplus.[52]

While a ban on MTBE would seem to have positive implications for ethanol producers, it could actually work against them. Because MTBE is more commonly used in RFG and high-octane gasoline, and because current ethanol production can not currently meet total U.S. demand for oxygenates and octane, there is also a push to suspend the oxygenate requirement in RFG, which would remove a major stimulus to the use of fuel ethanol. Furthermore, environmental groups and state air quality officials, although supportive of a ban on MTBE, are concerned over the possibility of "backsliding" if the oxygenate standard is eliminated. Because current RFG formulations have a lower level of toxic substances than is required under the Clean Air Act, there are concerns that new RFG formulations without oxygenates will meet the existing standard, but not the current level of overcompliance.

On March 20, 2000, the Clinton Administration announced a plan to reduce or eliminate MTBE use, and to promote the use of ethanol. Although no legislative language was suggested, the framework included three recommendations. The first was to "provide the authority to significantly reduce or eliminate the use of MTBE." The second recommendation was that "Congress must ensure that air quality gains are not diminished." The third was that "Congress should replace the existing oxygenate requirement in the Clean Air Act with a renewable fuel standard for all gasoline." Moreover, the Clinton Administration discussed the possibility of limiting the use of MTBE through the Toxic Substances Control Act (P.L. 94-469), which gives EPA the authority to control any substance that poses unreasonable risk to health or the environment. However, this process could take several years.[53] MTBE producers argued that such an initiative will lower clean air standards, and raise gasoline prices, while ethanol producers and some environmental groups were generally supportive of the announcement.[54]

In the 107th Congress, six MTBE-related bills have been introduced. (See Appendix 1.) All have been referred to Committee. These bills address different facets of the MTBE issue, including limiting or banning the use of MTBE, granting waivers to the oxygenate requirement, and authorizing funding for MTBE cleanup.

Alcohol Fuel Tax Incentives[55]

As stated above, the exemption that ethanol-blended fuels receive from the excise tax on motor fuels is controversial. The incentive allows fuel ethanol to compete with other additives, since the wholesale price of ethanol is so high. Proponents of ethanol argue that this exemption lowers dependence on foreign imports, promotes air quality, and benefits farmers.[56] A related, albeit smaller incentive for ethanol production is the small ethanol producers tax credit. This credit provides 10 cents per gallon for up to 15 million gallons of annual production by a small producer.[57]

Opponents of the tax incentives argue that the incentives promote an industry that could not exist on its own, and reduce potential fuel tax revenue. Despite objections from opponents, Congress in 1998 extended the motor fuels tax exemption through 2007, but at slightly lower rates (P.L. 105-178). A bill in the 107th Congress, S. 312, would increase the size of a covered producer under the small producer tax credit. (See Appendix 1.)

CONCLUSION

As a result of the current debate over the future of MTBE in RFG, and the RFG program in general, the future of the U.S. ethanol industry is uncertain. A ban on MTBE would greatly expand the market for ethanol, while an elimination of the oxygenate requirement would remove a major stimulus for its use. Any changes in the demand for ethanol will have major effects on corn producers, who rely on the industry as a partial market for their products.

The current size of the ethanol industry is depends significantly on federal laws and regulations that promote its use for air quality and energy security purposes, as well as tax incentives that lessen its cost to consumers. Without these, it is likely that the industry would shrink substantially in the near future. However, if fuel ethanol process costs can be decreased, or if gasoline prices increase, ethanol could increase its role in U.S. fuel consumption.

REFERENCES

[1]   Industrial uses include perfumes, aftershaves, and cleansers.
[2]   U.S. Department of Energy (DOE), Energy Information Administration (EIA). *Alternatives to Traditional Transportation Fuels* 1998. October 1999.
[3]   One bushel of corn generates approximately 2.5 gallons of ethanol.
[4]   Utilization data are used, rather than production, due to the existence of carryover stocks. Corn utilization data address the total amount of corn used within a given period.
[5]   Detailed explanations are available in *CRS Report* RS20271, Grain, Cotton, and Oilseeds: Federal Commodity Support, and *CRS* 98-744, Agricultural Marketing Assistance Loans and Loan Deficiency Payments.
[6]   The byproduct of the dry milling process is distillers dried grains. The byproducts of wet milling are corn gluten feed, corn gluten meal, and corn oil. Distillers dried grains, corn gluten feed, and corn gluten meal are used as livestock feed.
[7]   Renewable Fuels Association, *Ethanol Industry Outlook* 2001, Clean Air, Clean Water, Clean Fuel.
[8]   James Bovard, Archer Daniels Midland: A Case Study in Corporate Welfare. *Cato Institute*. September 26, 1995.
[9]   DOE, EIA, *Annual Energy Outlook* 2001. December 22, 2000. Table 18.
[10]   DOE, EIA, Alternatives to Traditional Transportation Fuels 1998. October 1999. Table 10.
[11]   DOE, EIA, *Annual Energy Outlook* 2001. December 22, 2000. Tables 2 and 18.
[12]   It should be noted that the use of ethanol does not effect the efficiency of an engine. There is simply less energy in one gallon of ethanol than in one gallon of gasoline.

[13] Section 211, subsections k and m (respectively). 42 U.S.C. 7545.

[14] CO, VOCs and nitrogen oxides (NOX)are the main precursors to ground-level ozone.

[15] Lead was commonly used as an octane enhancer until it was phased-out through the mid-1980s (lead in gasoline was completely banned in 1995), due to the fact that it disables emissions control devices, and because it is toxic to humans.

[16] P.L. 102-486.

[17] More diluted blends of ethanol, such as E10, are considered to be "extenders" of gasoline, as opposed to alternatives.

[18] Since different fuels produce different amounts of energy per gallon when consumed, the unit of a gasoline-equivalent gallon (GEG) is used to compare total energy consumption.

[19] DOE, EIA, *Alternatives to Traditional Transportation Fuels* 1998.

[20] Ibid.

[21] In 1997, some manufacturers made flexible E85/gasoline fueling capability standard on some models. It is expected, however, that most of these vehicles will be fueled by gasoline.

[22] Stacy C. Davis, DOE, *Transportation Energy Data Book:* Edition 20. November 2000.

[23] Executive Order 13134. August 12, 1999.

[24] 26 U.S.C. 40.

[25] *Food and Agriculture Policy Research Institute.* Effects on Agriculture of Elimination of the Excise Tax Exemption for Fuel Ethanol, Working Paper 01-97, April 8, 1997.

[26] James Bovard. p. 8.

[27] U.S. General Accounting Office, Effects of the Alcohol Fuels Tax Incentives. March, 1997.

[28] Ibid.

[29] Katrin Olson, "USDA Shows Losses Associated with Eliminating Ethanol Incentive," *Oxy-Fuel News.* May 19, 1997. p. 3.

[30] *Clean Air Act*, Section 211, subsection k. 42 U.S.C. 7545.

[31] Ground-level ozone is an air pollutant that causes smog, adversely affects health, and injures plants. It should not be confused with stratospheric ozone, which is a natural layer some 6 to 20 miles above the earth and provides a degree of protection from harmful radiation.

[32] Only the Los Angeles and New York areas are subject to both programs.

[33] *Clean Air Act*, Section 211, subsection m. 42 U.S.C. 7545.

[34] Margo T. Oge, Director, *Office of Mobile Sources*, U.S. EPA, Testimony Before the Subcommittee on Energy and Environment of the Committee on Science, U.S. House of Representatives. September 14, 1999.

[35] National Research Council, Ozone-Forming Potential of Reformulated Gasoline. May, 1999.

[36] Al Jessel, Senior Fuels Regulatory Specialist of Chevron Products Company, Testimony Before the House Science Committee Subcommittee on Energy and Environment. September 30, 1999.

[37] In 1998, an average of 90.9 million gallons per day of RFG were sold in the U.S., as opposed to 8.0 million gallons per day of Oxy-Fuel gasoline.

[38] California Energy Commission, Ethanol-Powered Vehicles.

[39] The fuel-cycle consists of all inputs and processes involved in the development, delivery and final use of the fuel.

[40] For example, nitrous oxide emissions tend to increase with ethanol use because nitrogen-based fertilizers are used extensively in agricultural production.

[41] M. Wang, C. Saricks, and D. Santini, "Effects of Fuel Ethanol on Fuel-Cycle Energy and Greenhouse Gas Emissions." Argonne National Laboratory.

[42] Hosein Shapouri, James A. Duffield, and Michael S. Graboski, USDA, Economic Researc Service, Estimating the Net Energy Balance of Corn Ethanol. July 1995.

[43] Wang, et. al. p. 1

[44] Shapouri, et. al. Table 1.

[45] U.S. General Accounting Office, Effects of the Alcohol Fuels Tax Incentives. March, 1997.

[46] Volatility of gasoline is its tendency to evaporate.

[47] RBOB: Reformulated Gasoline Blendstock for Oxygenate Blending.

[48] Estimates from the Renewable Fuels Association and EPA, respectively.

[49] Pamela Najer, "Refiners Get Flexibility to Blend Ethanol for Summer Fuel Supply in Two Cities," Daily Environment Report. March 19, 2001. p. A9.

[50] For more information, see CRS Report 98-290 ENR, MTBE in Gasoline: Clean Air and Drinking Water Issues.

[51] Blue Ribbon Panel on Oxygenates in Gasoline, Achieving Clean Air and Clean Water: The Report of the Blue Ribbon Panel on Oxygenates in Gasoline.

[52] Adrian Schofield, "Brazilian Ambassador Sees Opportunity in United States Ethanol Market," New Fuels & Vehicles Report. September 16, 1999. p. 1.

[53] U.S. Environmental Protection Agency, Headquarters Press Release: Clinton-Gore Administration Acts to Eliminate MTBE, Boost Ethanol. March 20, 2000.

[54] Jim Kennett, "Government Seeks to Ban Gas Additive," Houston Chronicle. March 21, 2000. p. A1

[55] For more information, see CRS Report 98-435E, Alcohol Fuels Tax Incentives.

[56] U.S. General Accounting Office (GAO), Effects of the Alcohol Fuels Tax Incentives. March, 1997.

[57] Defined as having a production capacity of less than 30 million gallons per year.

In: Biofuels in the Energy Supply System   ISBN: 1-59454-756-4
Editor: Victor I. Welborne, pp. 99-109 © 2006 Nova Science Publishers, Inc.

GLOSSARY

A

Acetic Acid — An acid with the structure of C2H4O2. Acetyl groups are bound through an ester linkage to hemicellulose chains, especially xylans, in wood and other plants. The natural moisture present in plants hydrolyzes the acetyl groups to acetic acid, particularly at elevated temperatures.

Acid Detergent Fiber (ADF) — organic matter that is not solubilized after 1 hour of refluxing in an acid detergent of cetyltrimethylammonium bromide in 1N sulfuric acid. ADF includes cellulose and lignin. This analytical method is commonly used in the feed and fiber industries. [8]

Acid Hydrolysis — The treatment of cellulosic, starch, or hemicellulosic materials using acid solutions (usually mineral acids) to break down the polysaccharides to simple sugars.

Acid Insoluble Lignin — Lignin is mostly insoluble in mineral acids, and therefore can be analyzed gravimetrically after hydrolyzing the cellulose and hemicellulose fractions of the biomass with sulfuric acid. ASTM E-1721-95 describes the standard method for determining acid insoluble lignin in biomass. See lignin and acid soluble lignin.

Acid Soluble Lignin — A small fraction of the lignin in a biomass sample is solubilized during the hydrolysis process of the acid insoluble lignin method. This lignin fraction is referred to as acid soluble lignin and may be quantified by ultraviolet spectroscopy [1]. See lignin and acid insoluble lignin.

Agricultural Residue — Agricultural crop residues are the plant parts, primarily stalks and leaves, not removed from the fields with the primary food or fiber product. Examples include corn stover (stalks, leaves, husks, and cobs); wheat straw; and rice straw. With approximately 80 million acres of corn planted annually, corn stover is expected to become a major biomass resource for bioenergy applications.

Aldoses — Occur when the carbonyl group of a monosaccharide is an aldehyde [2].

Alkali Lignin — Lignin obtained by acidification of an alkaline extract of wood.

Aquatic Plants — The wide variety of aquatic biomass resources, such as algae, giant kelp, other seaweed, and water hyacinth. Certain microalgae can produce hydrogen and oxygen while others manufacture hydrocarbons and a host of other products. Microalgae examples include Chlorella, Dunaliella, and Euglena.

Arabinan — The polymer of arabinose with a repeating unit of C5H804 [2]. Can be hydrolyzed to arabinose.

Arabinose — A five-carbon sugar C5H1005. A product of hydrolysis of arabinan found in the hemicellulose fraction of biomass.

Ash — Residue remaining after ignition of a sample determined by a definite prescribed procedure [3].

B

Bark — The outer protective layer of a tree outside the cambium comprising the inner bark and the outer bark. The inner bark is a layer of living bark that separates the outer bark from the cambium and in a living tree is generally soft and moist. The outer bark is a layer of dead bark that forms the exterior surface of the tree stem. The outer bark is frequently dry and corky. [8]

Biomass — Any plant-derived organic matter. Biomass available for energy on a sustainable basis includes herbaceous and woody energy crops, agricultural food and feed crops, agricultural crop wastes and residues, wood wastes and residues, aquatic plants, and other waste materials including some municipal wastes. Biomass is a very heterogeneous and chemically complex renewable resource.

Biomass Processing Residues — Byproducts from processing all forms of biomass that have significant energy potential. For example, making solid wood products and pulp from logs produces bark, shavings and sawdust, and spent pulping liquors. Because these residues are already collected at the point of processing, they can be convenient and relatively inexpensive sources of biomass for energy.

C

Carbohydrate — Organic compounds made up of carbon, hydrogen, and oxygen and having approximately the formula $(CH2O)$ n; includes cellulosics, starches, and sugars [8].

Cellulose — The carbohydrate that is the principal constituent of wood and other biomass and forms the structural framework of the wood cells. It is a polymer of glucose with a repeating unit of $C6H10O5$ strung together by ß-glycosidic linkages. The ß-linkages in cellulose form linear chains that are highly stable and resistant to chemical attack because of the high degree of hydrogen bonding that can occur between chains of cellulose (see below). Hydrogen bonding between cellulose chains makes the polymers more rigid, inhibiting the flexing of the molecules that must occur in the hydrolytic breaking of the glycosidic linkages. Hydrolysis can reduce cellulose to a cellobiose repeating unit, $C12H22O11$, and ultimately to glucose, $C6H12O6$. Heating values for cellulose may be slightly different based upon the feedstock. Example values are shown below (higher heating value [HHV] at 30°C, dry basis) [3].

cotton linters: HHV=7497 BTU/LB (4172.0 cal/g, 17426.4 J/g)

wood pulp: HHV=7509.6 BTU/LB (4165.0 cal/g, 17455.6 J/g)

Linear Chains of Glucose linked by b-Glycosidic Bonds Comprise Cellulose

Linear Chains of Glucose linked by b-Glycosidic Bonds Comprise Cellulose

Chips — small fragments of wood chopped or broken by mechanical equipment. Total tree chips include wood, bark, and foliage. Pulp chips or clean chips are free of bark and foliage.

E

Elemental Analysis — The determination of carbon, hydrogen, nitrogen, oxygen, sulfur, chlorine and ash in a sample. See Ultimate Analysis.

Extractives — Any number of different compounds in biomass that are not an integral part of the cellular structure. The compounds can be extracted from wood by means of polar and non-polar solvents including hot or cold water, ether , benzene, methanol, or other solvents that do not degrade the biomass structure. The types of extractives found in biomass samples are entirely dependent upon the sample itself [4].

F

Fixed Carbon — The carbon remaining after heating in a prescribed manner to decompose thermally unstable components and to distill volatiles. Part of the proximate analysis group.

Forestry Residues — Includes tops, limbs, and other woody material not removed in forest harvesting operations in commercial hardwood and softwood stands, as well as woody material resulting from forest management operations such as precommercial thinnings and removal of dead and dying trees.

G

Galactan — The polymer of galactose with a repeating unit of $C6H10O5$. Found in hemicellulose it can be hydrolyzed to galactose.

Galactose — A six-carbon sugar with the formula $C6H12O6$. A product of hydrolysis of galactan found in the hemicellulose fraction of biomass.

Glucan — The polymer of glucose with a repeating unit of $C6H10O5$ [2]. Cellulose is a form of glucan. Can be hydrolyzed to glucose.

Glucose — A simple six-carbon sugar $C6H12O6$. A product of hydrolysis of glucan found in cellulose and starch. A sweet, colorless sugar that is the most common sugar in nature and the sugar most commonly fermented to ethanol.

Guaiacyl — A chemical component of lignin. It has a six-carbon aromatic ring with one methoxyl group attached. It is the predominant aromatic structure in softwood lignins. See syringyl.

H

Hardwood — One of the botanical groups of dicotyledonous trees that have broad leaves in contrast to the conifers or softwoods. The term has no reference to the actual hardness of the wood. The botanical name for hardwoods is angiosperms. Short-rotation, fast growing hardwood trees are being developed as future energy crops. They are uniquely developed for harvest from 5 - 8 years after planting. Examples include: Hybrid poplars (Populus sp.), Hybrid willows (Salix sp.), Silver maple (Acer saccharinum), and Black locust (Robinia pseudoacacia).

Heating Value — Higher heating value (HHV) is the potential combustion energy when water vapor from combustion is condensed to recover the latent heat of vaporization. Lower heating value (LHV) is the potential combustion energy when water vapor from combustion is not condensed. See also higher heating value and lower heating value.

Hemicellulose — Hemicellulose consists of short, highly branched chains of sugars. In contrast to cellulose, which is a polymer of only glucose, a hemicellulose is a polymer of five different sugars. It contains five-carbon sugars (usually D-xylose and L-arabinose) and six-carbon sugars (D-galactose, D-glucose, and D-mannose) and uronic acid. The sugars are highly substituted with acetic acid. The branched nature of hemicellulose renders it amorphous and relatively easy to hydrolyze to its constituent sugars compared to cellulose. When hydrolyzed, the hemicellulose from hardwoods releases products high in xylose (a five-carbon sugar). The hemicellulose contained in softwoods, by contrast, yields more six-carbon sugars.

Herbaceous Plants — Non-woody species of vegetation, usually of low lignin content such as grasses.

Herbaceous Plants — Non-woody species of vegetation, usually of low lignin content such as grasses.

Herbaceous Energy Crops — Perennial non-woody crops that are harvested annually, though they may take 2 to 3 years to reach full productivity. Examples include: Switchgrass (Panicum virgatum), Reed canarygrass (Phalaris arundinacea), Miscanthus (Miscanthus x giganteus), and Giant reed (Arundo donax).

Hexose — any of various simple sugars that have six carbon atoms per molecule (e.g. glucose, mannose, and galactose.)

Higher Heating Value (HHV, also known as Gross Heat of Combustion) — The heat produced by combustion of one unit of substance at constant volume in an oxygen bomb calorimeter under specified conditions. The conditions are: initial oxygen pressure of 2.0-4.0 MPa (20-40 atm), final temperature of 20°-35°C, products in the form of ash, liquid water, gaseous CO_2 and N_2, and dilute aqueous HCl and H_2SO_4. It is assumed that if significant quantities of metallic elements are combusted, they are converted to their oxides. In the case of materials such as coal, wood, or refuse, if small or trace amounts of metallic elements are present, they are unchanged during combustion and are part of the ash.

Holocellulose — The total carbohydrate fraction of wood — cellulose plus hemicellulose.

Hydrolysis — The conversion, by reaction with water, of a complex substance into two or more smaller units, such as the conversion of cellulose into glucose sugar units.

K

Klason Lignin — Lignin obtained from wood after the non-lignin components of the wood have been removed with a prescribed sulfuric acid treatment. A specific type of acid-insoluble lignin analysis.

L

Lignin — The major noncarbohydrate, polypenolic structural constituent of wood and other native plant material that encrusts the cell walls and cements the cells together. It is a highly polymeric substance, with a complex, cross-linked, highly aromatic structure of molecular weight about 10,000 derived principally from coniferyl alcohol ($C_{10}H_{12}O_3$) by extensive condensation polymerization. Higher heating value (oven dry basis): HHV=9111 BTU/LB (5062 CAL/G, 21178 J/G) [3].

Lignin Ratio of MeO to C9 — Lignin empirical formulae are based on ratios of methoxy groups to phenylpropanoid groups (MeO:C9). The general empirical formula for lignin monomers is $C_9H_{10}O_2$ (OCH_3)n, where n is the ratio of MeO to C9 groups. Where no experimental ratios have been found, they are estimated as follows: 0.94 for softwoods; 1.18 for grasses; 1.4 for hardwoods. These are averages of the lignin ratios found in the literature. Paper products, which are produced primarily from softwoods, are estimated to have an MeO:C9 ratio of 0.94.

Lignin Pseudo-Molecule for Modeling — The n Lignin ratio of methoxy groups to phenylpropanoid groups (MeO:C9) is used to calculate an ultimate analysis for the lignin pseudo-molecule and then this ultimate analysis is used to estimate other properties of the molecule, such as its higher and lower heating values.

Lignocellulose — Refers to plant materials made up primarily of lignin, cellulose, and hemicellulose.

Lower Heating Value (LLV also known as Net Heat of Combustion) — The heat produced by combustion of one unit of a substance, at atmospheric pressure under conditions such that all water in the products remains in the form of vapor. The net heat of combustion is calculated from the gross heat of combustion at 20°C by subtracting 572 cal/g (1030 Btu/lb) of water derived from one unit mass of sample, including both the water originally present as moisture and that formed by combustion. This subtracted amount is not equal to the latent heat of vaporization of water because the calculation also reduces the data from the gross value at constant volume to the net value at constant pressure. The appropriate factor for this reduction is 572 cal/g [3].

M

Mannan — The polymer of mannose with a repeating unit of C6H10O5 [2]. Can be hydrolyzed to mannose.

Mannose — A six-carbon sugar C6H12O6. A product of hydrolysis of mannan found in the hemicellulose fraction of biomass.

Mass Closure (%) — The percent by weight of the total samples extracted from the biomass sample compared to the weight of the original sample. It is a sum of the weight percent of moisture, extractives, ash, protein, total lignin, acetic acid, uronic acids, arabinan, xylan, mannan, galactan, glucan, and starch. This is a good indicator of the accuracy of a complete biomass compositional analysis.

Moisture — This is a measure of the amount of water and other components that are volatilized at 105°C present in the biomass sample [6].

Moisture Free Basis — Biomass composition and chemical analysis data is typically reported on a moisture free or dry weight basis. Moisture (and some volatile matter) is removed prior to analytical testing by heating the sample at 105°C to constant weight. By definition, samples dried in this manner are considered moisture free.

Monosaccharide — a simple sugar such as a five-carbon sugar (xylose, arabinose) or six-carbon sugar (glucose, fructose). Sucrose, on the other hand is a disaccharide, composed of a combination of two simple sugar units, glucose and fructose.

Municipal Wastes — Residential, commercial, and institutional post-consumer wastes contain a significant proportion of plant-derived organic material that constitutes a renewable energy resource. Waste paper, cardboard, construction and demolition wood waste, and yard wastes are examples of biomass resources in municipal wastes.

N

Native Lignin — The lignin as it exists in the lignocellulosic complex before separation.

Neutral Detergent Fiber (NDF) — Organic matter that is not solubilized after one hour of refluxing in a neutral detergent consisting of sodium lauryl sulfate and EDTA at pH 7. NDF includes hemicellulose, cellulose, and lignin. [8]

P

Polysaccharide — A long-chain carbohydrate containing at least three molecules of simple anhydrosugars linked together. Examples include cellulose and starch.

Proximate Analysis — The determination, by prescribed methods, of moisture, volatile matter, fixed carbon (by difference), and ash. The term proximate analysis does not include determinations of chemical elements or determinations other than those named [7]. The group of analyses is defined in ASTM D 3172.

Protein — A protein molecule is a chain of up to several hundred amino acids and is folded into a more or less compact structure. Because about 20 different amino acids are used by living matter in making proteins, the variety of protein types is enormous. In their biologically active states, proteins function as catalysts in metabolism and to some extent as structural elements of cells and tissues [7]. Protein content in biomass (in mass %) can be estimated by multiplying the mass % nitrogen of the sample by 6.25.

R

Residues, Biomass — Byproducts from processing all forms of biomass that have significant energy potential. For example, making solid wood products and pulp from logs produces bark, shavings and sawdust, and spent pulping liquors. Because these residues are already collected at the point of processing, they can be convenient and relatively inexpensive sources of biomass for energy.

S

Saccharide — A simple sugar or a more complex compound that can be hydrolyzed to simple sugar units.

Softwood — Generally, one of the botanical groups of trees that in most cases have needle-like or scale-like leaves; the conifers; also the wood produced by such trees. The term has no reference to the actual hardness of the wood. The botanical name for softwoods is gymnosperms.

Starch — A molecule composed of long chains of a-glucose molecules linked together (repeating unit C12H16O5). These linkages occur in chains of a-1,4 linkages with branches formed as a result of a-1,6 linkages (see below). This polysaccharide is widely distributed in the vegetable kingdom and is stored in all grains and tubers. A not-so-obvious consequence of the a linkages in starch is that this polymer is highly amorphous, making it more readily attacked by human and animal enzyme systems and broken down into glucose. Gross heat of combustion: Qv(gross)=7560 Btu/lb (4200 cal/g,17570 J/g)[3].

Diagram of the structure of glucose

The polymeric structure of glucose in starch tends to be amorphous

Stover — The dried stalks and leaves of a crop remaining after the grain has been harvested.

Structural Chemical Analysis — The composition of biomass reported by the proportions of the major structural components; cellulose, hemicellulose, and lignin. Typical ranges are shown in the table below.

Component	Percent Dry Weight
Cellulose	40-60%
Hemicellulose	20-40%
Lignin	10-25%

Syringyl — A component of lignin, normally only found in hardwood lignins. It has a six-carbon aromatic ring with two methoxyl groups attached. See guaiacyl.

T

Total Lignin — The sum of the acid soluble lignin and acid insoluble lignin fractions.

Total Solids — The amount of solids remaining after all volatile matter has been removed from a biomass sample by heating at 105°C to constant weight [6].

U

Ultimate Analysis — The determination of the elemental composition of the organic portion of carbonaceous materials, as well as the total ash and moisture. Determined by prescribed methods. See elemental analysis [8].

Uronic Acid — A simple sugar whose terminal -CH2OH group has been oxidized to an acid, COOH group. The uronic acids occur as branching groups bonded to hemicelluloses such as xylan [8].

V

Volatile Matter — Those products, exclusive of moisture, given off by a material as a gas or vapor, determined by definite prescribed methods that may vary according to the nature of the material [8]. One definition of volatile matter is part of the proximate analysis group usually determined as described in ASTM D 3175.

W

Whole Tree Chips — wood chips produced by chipping whole trees, usually in the forest. Thus the chips contain both bark and wood. They are frequently produced from the low-quality trees or from tops, limbs, and other logging residues.

Willstatter Lignin — Lignin obtained from the lignocellulosic complex after it has been extracted with fuming hydrochloric acid.

Wood — a solid lignocellulosic material naturally produced in trees and some shrubs, made of up to 40%-50% cellulose, 20%-30% hemicellulose, and 20% -30% lignin.

X

Xylan — A polymer of xylose with a repeating unit of C5H804, found in the hemicellulose fraction of biomass. Can be hydrolyzed to xylose. Gross heat of combustion: Qv(gross)=17751.9 Jg-1[3].

Xylose — A five-carbon sugar C5H1005. A product of hydrolysis of xylan found in the hemicellulose fraction of biomass.

INDEX

Y